An Introduction to the Theory of CANONICAL MATRICES

An Introduction to the Theory of CANONICAL MATRICES

H. W. TURNBULL, M.A., F.R.S.

Formerly Regius Professor of Mathematics in the United College, University of St. Andrews

A. C. AITKEN, D.Sc., F.R.S.

Professor of Mathematics in the University of Edinburgh

DOVER PUBLICATIONS, INC.
Mineola, New York

DOVER PHOENIX EDITIONS

Bibliographical Note

This Dover edition, first published in 1961 and reissued in 2005, is an unabridged and corrected republication of the third (1952) edition of the work first published by Blackie & Son Limited, London and Glasgow, in 1932.

International Standard Book Number: 0-486-44168-7

Manufactured in the United States of America
Dover Publications, Inc., 31 East 2nd Street, Mineola, N.Y. 11501

PREFACE

This book has been written with the object of giving an account of the various ways in which matrices of finite order can be reduced to canonical form under different important types of transformation. While the work has been planned to serve as a sequel to a former publication, *The Theory of Determinants, Matrices, and Invariants* (1928), circumstances have allowed us to make it practically independent and self-contained, with the least possible overlapping of material in the two books. A certain knowledge of the elementary theory of determinants is presupposed, but no previous acquaintance with matrices.

The volume on *Invariants*—as it will be referred to in subsequent pages—in giving an introductory account of matrices and determinants, treated only of such properties as belonged to the general linear transformation; for these are the properties which have the most direct bearing on the projective invariant theory, to which the later chapters were devoted. In the nomenclature of the work before us, the treatment was confined to the diagonal case of the classical canonical form, in which the elementary divisors are necessarily linear.

In the present work we return to consider, in close detail, those important cases in which the elementary divisors are no longer restricted to be linear, but may be of general degree. To adopt a geometrical mode of speaking, it is as if we had formerly been concerned purely with the projective properties of quadrics in general position, but had now returned to the consideration of all possible distinctions between quadrics under certain prescribed conditions; such distinctions, for example, as those which persist through all projective transformations, or again through all rotations, and so on.

The subject-matter of the canonical reduction of matrices, which has numerous and important applications, has received attention in several treatises and a large number of original papers. The historical notes which we have appended to each chapter are intended to give a brief review of what has been done on each topic, to apportion due credit to pioneers, and to stimulate the student to further reading. (We would warn him, however, to make sure at the outset, in reading any work on groups or matrices, whether the author means AB or what we have denoted by BA when he writes a product.) The most complete accounts of the theory available are those of Muth (*Elementarteiler*, 1899) and Cullis (*Matrices and Determinoids*, Vols. I, II, III, 1913, 1918, 1925). We have preferred to follow the lead of Cullis, who develops the theory in terms of the structure and properties of matrices—in matrix idiom, as it were, rather than in terms of bilinear and quadratic forms, or of linear substitutions.

We take the opportunity of acknowledging our indebtedness to the work of those writers who have given a sustained account of the theory, in one guise or another; in particular to Muth, as above, to Bromwich (Cambridge Tract on *Quadratic Forms*, 1906, and various papers), to Bôcher (*Higher Algebra*, 1907), Hilton (*Linear Substitutions*, 1914), Cullis (Vol. III of *Matrices and Determinoids*, 1925), and Dickson (*Modern Algebraic Theories*, 1926).

While we have tried to include all the principal features of the theory and have sought to make the sequence of argument reasonably fluent, even allowing ourselves moderate latitude in digression and explanation, we have, at the same time, aimed at a certain compactness in the formulæ and demonstrations. This has been achieved in the first place by a systematic use of the matrix notation, to which we shall again refer; in the second place, by confining the contents of each chapter almost entirely to general theorems, and by relegating corollaries and applications to the interspersed sets of examples. These examples are intended to serve not so much as exercises, many being quite easy, but rather as points of relaxation, and running commentary; they will, however, be found to contain many well-known and important theorems, which the notation establishes in the minimum of space.

We attach the greatest importance to the choice of notation. Inferring from perusal of Cullis that the emphasis laid since the time of Cayley on the square matrix might well be removed, we resolved to continue the plan adopted in *Invariants* by making the fullest use of rectangular matrices and submatrices, and of partitioned matrices, by insisting on the condition that the non-commutative rules of product order hold without exception, and by distinguishing always between a matrix of a single row and one of a single column. When this is done, all the systems which appear, whether scalars, vectors, or matrices, can be regarded as rectangular matrices or products of rectangular matrices, and the theory is thus greatly unified. We would draw special attention to the notation $x'Ay$ for the bilinear form, $x'Ax$ for the quadratic, and $\bar{x}'Ax$ for the Hermitian form, believing that these notations will enable the linear transformations and the bilinear, quadratic, and Hermitian forms which are fundamental, for example, in analytical geometry, dynamics, or mathematical statistics, to be manipulated with ease.

Through considerations of space we have not been able to include many applications to geometry, but the results are readily adaptable: nor to the theory of Groups, where, as Schur has shown, partitioned matrices can be used with elegance and advantage.

The reader already familiar with the theory will also observe that certain established methods of dealing with the subject have hardly been touched upon, notably the methods of Weierstrass and Darboux, the theory of regular minors of determinants and the treatment of quadratic forms by the methods of Kronecker. We have, in fact, allowed ourselves a free hand in dealing with the results of earlier writers, in the belief that the outcome would prove to be an easier approach to a subject that has often failed to win affection; and the methods of H. J. S. Smith, Sylvester, Frobenius, and Dickson proved in themselves quite adequate without the inclusion of other parallel theories. A thorough assimilation of the algebraic implications of Euclid's H.C.F. process, and of the notion of linear dependence, furnishes the clue to many passages. Our tribute to Kronecker finds expression in Chapter IX, which is an essay towards giving a fresh derivation of his classical results concerning singular pencils; we have treated this by rational methods, and we trust that an intricate argument has been materially simplified.

Our best thanks are due to Dr. E. T. Copson and Mr. D. E. Rutherford at St. Andrews University, who have taken an interest in the progress of the work, and have offered valuable suggestions at the proof-reading stage; and especially to Dr. John Dougall, for his critical vigilance and expert mathematical and technical help during the passage of the work through the press.

H. W. T.
A. C. A.

ST. ANDREWS / EDINBURGH — *December*, 1931.

PREFACE TO THE SECOND IMPRESSION

Various emendations of the text of the First Edition have been incorporated in the present impression. We take occasion to thank those friends who first brought the principal ones to our notice: in particular Dr. J. Williamson, who pointed out the gap in the theory of Hermitian pencils (p. 131) now supplied by the signature test, and Dr. W. Ledermann, who gave valuable criticism on the subject of singular pencils.

H. W. T.
A. C. A.

St. Andrews
Edinburgh } *November*, 1944.

PREFACE TO THE THIRD IMPRESSION

The book has been extended by an Appendix which, as continuing the theory by the rational methods of reduction, may be read as a supplement to Chapters V and X. A few additional examples have been inserted, particularly on p. 30 and p. 57. Our thanks are more especially due to Dr. D. E. Rutherford, whose writings have suggested the material of the Appendix.

H. W. T.
A. C. A.

Millom
Edinburgh } *October*, 1951.

CONTENTS

CHAPTER I

DEFINITIONS AND FUNDAMENTAL PROPERTIES OF MATRICES

CHAPTER II

ELEMENTARY TRANSFORMATIONS. BILINEAR AND QUADRATIC FORMS

CHAPTER III

THE CANONICAL REDUCTION OF EQUIVALENT MATRICES

Chapter IV

SUBGROUPS OF THE GROUP OF EQUIVALENT TRANSFORMATIONS

Chapter V

A RATIONAL CANONICAL FORM FOR THE COLLINEATORY GROUP

Chapter VI

THE CLASSICAL CANONICAL FORM FOR THE COLLINEATORY GROUP

Chapter VII

CONGRUENT AND CONJUNCTIVE TRANSFORMATIONS: QUADRATIC AND HERMITIAN FORMS

Chapter VIII

CANONICAL REDUCTION BY UNITARY AND ORTHOGONAL TRANSFORMATIONS

CHAPTER IX

THE CANONICAL REDUCTION OF PENCILS OF MATRICES

CHAPTER X

APPLICATIONS OF CANONICAL FORMS TO SOLUTION OF LINEAR MATRIX EQUATIONS. COMMUTANTS AND INVARIANTS

Chapter XI

PRACTICAL APPLICATIONS OF CANONICAL REDUCTION

The
Theory of Canonical Matrices

CHAPTER I

Definitions and Fundamental Properties of Matrices

1. Introductory.

The theory of canonical matrices is concerned with the systematic investigation of types of transformation which reduce matrices to the simplest and most convenient shape. The formulation of these various types is not merely useful as a preliminary to the deeper study of the properties of matrices themselves; it serves also to render the theory of matrices more immediately available for numerous applications to geometry, differential equations, analytical dynamics, and the like. Quite early, for example, in co-ordinate geometry, when the equation of a general conic is simplified by reference to principal axes, or again when two general conics are referred to their common self-conjugate triangle, the procedure involved is really equivalent to the canonical reduction of a matrix.

2. Definitions and Fundamental Properties.

It will be of advantage to recall briefly the definitions and fundamental properties of matrices. By a matrix A of order n is meant a system of elements, which may be real or complex numbers, arranged in a square formation of n rows and columns,

$$A = [a_{ij}] = \begin{bmatrix} a_{11} & a_{12} & \dots & a_{1n} \\ a_{21} & a_{22} & \dots & a_{2n} \\ \cdot & \cdot & \cdot & \cdot \\ \cdot & \cdot & \cdot & \cdot \\ a_{n1} & a_{n2} & \dots & a_{nn} \end{bmatrix}, \qquad \dots \quad (1)$$

where a_{ij} denotes the element standing in the ith row and the jth column, the (ij)th element, as we shall frequently call it. The determinant having the same elements is denoted by $|A|$, or $|a_{ij}|$, and is naturally called the determinant of the matrix A. We shall also make continual use of *rectangular* matrices, of m rows and n columns, or, as it will be phrased, of order m by n, $m \times n$. Where there is only one row, so that $m = 1$, such a matrix will be termed a *vector of the first kind*, or a *prime*; and it will often be denoted by a single small italic u or v. Thus

$$u = [u_1, u_2, \ldots, u_n]. \quad \ldots \ldots \quad (2)$$

On the other hand, a matrix of a single column, of n elements, will be termed a *vector of the second kind*, or a *point*; and to save space it will not be printed vertically but horizontally, and distinguished by brackets $\{\ldots\}$. Thus

$$x = \{x_1, x_2, \ldots, x_n\}. \quad \ldots \ldots \quad (3)$$

The accented matrix $A' = [a_{ji}]$, obtained by complete interchange of rows and columns in A, is called the *transpose* of A. The ith row $[a_{i1}, a_{i2}, \ldots, a_{in}]$ of A is identical with the ith column of A'. For vectors we have $u' = [u_j]' = \{u_j\}$, $x' = \{x_i\}' = [x_i]$.

Matrices may be multiplied either by ordinary *scalar* numbers or by matrices. The effect of multiplying a matrix $A = [a_{ij}]$ by a scalar λ is to multiply each element of A by λ. The product is defined by

$$\lambda A = \lambda[a_{ij}] = [\lambda a_{ij}] = A\lambda. \quad \ldots \ldots \quad (4)$$

Matrices of the same order are added, or subtracted, by adding, or subtracting, corresponding elements; so that a linear combination of two such matrices A and B, with scalar multipliers λ and μ, is defined by

$$\lambda A + \mu B = [(\lambda a_{ij} + \mu b_{ij})]. \quad \ldots \ldots \quad (5)$$

Hence, if $C = \lambda A + \mu B$, then $c_{ij} = \lambda a_{ij} + \mu b_{ij}$: and also $C' = \lambda A' + \mu B'$, for the transposed matrices.

The *null* or *zero matrix*, whether square or rectangular, has all its elements zero, and will often be denoted without ambiguity by an ordinary cipher. The *unit matrix*, I, is necessarily square; it has a unit for each element in the principal diagonal, and the remaining elements all zero. Thus

$$I = [\delta_{ij}], \qquad \delta_{ij}\begin{bmatrix} = 0, & i \neq j, \\ = 1, & i = j. \end{bmatrix} \quad \ldots \ldots \quad (6)$$

3. Matrix Multiplication.

The multiplication of matrices by matrices, or matrix multiplication, differs in important respects from scalar multiplication. Two matrices can be multiplied together only when the number of columns in the first is equal to the number of rows in the second. Matrices which satisfy this condition will be termed *conformable* matrices; their product AB is defined by

$$AB = [a_{ij}]\,[b_{ij}] = [\sum_{k=1}^{p} a_{ik}b_{kj}] = [c_{ij}] = C, \quad . \quad . \quad . \qquad (7)$$

where the orders of A, B, C are $m \times p$, $p \times n$, $m \times n$ respectively. The process of multiplication is thus the same as the *row-by-column* rule for multiplying together determinants of equal order. If the matrices are square and each of order n, then the corresponding relation $|A|\,|B| = |C|$ is true for the determinants $|A|$, $|B|$, $|C|$.

Matrices are regarded as equal only when they are element for element identical. Therefore, since a row-by-column rule will in general give different elements from a column-by-row rule, the product BA, if it exists at all, is usually different from AB. (AB and BA, it may be observed, can coexist only if $m = n$.) We must therefore distinguish always between *premultiplication*, as when B, premultiplied by A, yields the product AB, and *postmultiplication*, as when B, postmultiplied by A, yields the product BA. If $AB = BA$ the matrices A and B are said to *commute*, or to be *permutable*; and one of the applications of the theory of canonical matrices is to find the general matrix X permutable with a given matrix A. Except for the noncommutative law of multiplication (and therefore of division, defined as the inverse operation) all the ordinary laws of algebra apply to matrices, very much as they do in the elementary theory of vectors. Of particular importance is the associative law $(AB)C = A(BC)$, which allows us to dispense with brackets and to write ABC without ambiguity, since the double summation $\sum_k \sum_l a_{ik}b_{kl}c_{lj}$ can be carried out in either of the orders indicated. Similarly for the sum $A + B + C$.

The above remarks are restricted to the case of matrices of finite order; for the associative law of multiplication does not necessarily hold when any of the matrices involved has one or both of the orders m, n infinite.

The integers m, n, p which appear in (7) may take any positive value. One extreme case, when $m = n = 1$, yields the *inner product* of the vectors u and x. Thus

$$ux = u_1x_1 + u_2x_2 + \ldots + u_px_p = \sum_k u_kx_k = x'u'. \quad . \qquad (8)$$

The product here is a scalar. On the other hand, the product xu, which exemplifies (7) with $p=1$, $m=n$, is a square matrix of order n, having $x_i u_j$ for its (ij)th element, namely

$$xu = [x_i u_j] = (u'x')'. \quad . \quad . \quad . \quad . \quad . \quad . \quad (9)$$

4. Reciprocal of a Non-Singular Matrix.

When the determinant $|A| = |a_{ij}|$ of a square matrix A does not vanish A is said to be *non-singular*, and possesses a *reciprocal* or *inverse* matrix R such that

$$AR = RA = I.$$

The reciprocal R is unique, as will be seen, and is readily obtained, from the theory of determinants. If A_{ij} denotes the co-factor of a_{ij} in $|A|$, the matrix $[A_{ji}]$ is called the *adjoint* of A, and exists whether A is singular or not. (The determinant $|A_{ij}|$ is the *adjugate* of $|A|$.) It follows that

$$[a_{ij}]\,[A_{ji}] = [\sum_k a_{ik} A_{jk}] = [\,|A|\,\delta_{ij}] = |A|\,I. \quad . \quad (10)$$

Thus the product of A and its adjoint is that special type of diagonal matrix called a *scalar matrix*; each diagonal element $(i=j)$ is equal to the determinant $|A|$, and the rest are zero. If $|A| \neq 0$, we may divide throughout by the scalar $|A|$, obtaining at once the required form of R. The (ij)th element of R is therefore $A_{ji}\,|A|^{-1}$, or, let us say, a^{ji}, where the *reversed* order of upper indices must be carefully noted. Writing now A^{-1} instead of R, we have

$$A^{-1} = [a^{ji}] = [A_{ji}\,|A|^{-1}], \qquad |A| \neq 0. \quad . \quad . \quad (11)$$

By actual multiplication $AA^{-1} = A^{-1}A = I$; so that the name reciprocal and the notation A^{-1} are justified. It may be observed in passing that in products of matrices the unit factor I may be introduced or suppressed at pleasure, like the unit factor of scalar algebra.

5. The Reversal Law in Transposed and Reciprocal Products.

A fundamental consequence of the non-commutative law of matrix multiplication is the *reversal* law, exemplified in transposing and reciprocating a continued product of matrices. Thus

$$(AB)' = B'A', \;\; (ABC)' = C'B'A'; \quad . \quad . \quad . \quad (12)$$

and, if $|A| \neq 0$, $|B| \neq 0$, $|C| \neq 0$,

$$(AB)^{-1} = B^{-1}A^{-1}, \;\; (ABC)^{-1} = C^{-1}B^{-1}A^{-1}. \quad . \quad . \quad (13)$$

EXAMPLES

1. Prove that the reciprocal of a non-singular matrix is unique. [If $AR = I$, and also $AS = I$, then $AR - AS = 0$, the null matrix. By the distributive law $A(R - S) = 0$, and hence $A^{-1}A(R - S) = 0$. Thus $I(R - S) = R - S = 0$; and so $R = S$. All solutions X of the equation $AX = I$ are therefore equal. But A^{-1} is a solution and is therefore the unique solution.]

2. Verify (13) by premultiplying by B, A, or C, B, A in turn.

3. Prove that $[a^{ij}][b^{ij}] = [c^{ij}]$, where $i, j = 1, 2, \ldots, n$, provided that $c^{ij} = \sum_{k=1}^{n} a^{ik}b^{kj}$.

4. If A is a square matrix of order n, while u and x are vectors of the row and column kinds respectively, then uA denotes a row vector while Ax denotes a column vector. The products Au, xA are undefined if $n > 1$.

5. What do $A'u'$, $x'A'$ represent? [Column vector, row vector.]

6. Matrices partitioned into Submatrices.

It is convenient to extend the use of the fundamental laws of combination for matrices to the case where a matrix is regarded as constructed not so much from elements as from submatrices, or minor matrices, of elements. (Cf. *Invariants*, p. 38.) For example, the matrix

$$A = \left[\begin{array}{cc:c} 1 & 2 & 3 \\ 4 & 5 & 6 \\ \hdashline 7 & 8 & 9 \end{array}\right]$$

can be written

$$A = \begin{bmatrix} P & Q \\ R & S \end{bmatrix},$$

where

$$P = \begin{bmatrix} 1 & 2 \\ 4 & 5 \end{bmatrix}, \quad Q = \begin{bmatrix} 3 \\ 6 \end{bmatrix}, \quad R = [7, 8], \quad S = [9].$$

Here the diagonal submatrices P and S are square, and the partitioning is diagonally symmetrical. In the general case there may evidently be n or fewer partitions row-wise or column-wise. Let B be a second square matrix of the third order similarly partitioned:

$$B = \begin{bmatrix} P_1 & Q_1 \\ R_1 & S_1 \end{bmatrix} = \left[\begin{array}{cc:c} 2 & . & 1 \\ 3 & 1 & 2 \\ \hdashline 1 & 2 & . \end{array}\right]:$$

then by addition and multiplication we have

$$A+B = \begin{bmatrix} P+P_1 & Q+Q_1 \\ R+R_1 & S+S_1 \end{bmatrix}, \quad AB = \begin{bmatrix} PP_1+QR_1 & PQ_1+QS_1 \\ RP_1+SR_1 & RQ_1+SS_1 \end{bmatrix}, \tag{14}$$

as may readily be verified. In each case the resulting matrix is of the same order, and is partitioned in the same way, as the original matrix factors. For example, in AB the first element, $PP_1 + QR_1$, stands for a square submatrix of two rows and columns: and this is possible since, by definition, both products PP_1 and QR_1 consist of two rows and two columns. Similar remarks apply to the other submatrix "elements". Thus

$$PQ_1 + QS_1 = \begin{bmatrix} 1 & 2 \\ 4 & 5 \end{bmatrix} \begin{bmatrix} 1 \\ 2 \end{bmatrix} + \begin{bmatrix} 3 \\ 6 \end{bmatrix} [0] = \begin{bmatrix} 5 \\ 14 \end{bmatrix} + \begin{bmatrix} 0 \\ 0 \end{bmatrix} = \begin{bmatrix} 5 \\ 14 \end{bmatrix},$$

giving the proper rectangular shape for the upper right-hand minor.

It was observed earlier that a rectangular matrix B could be pre-multiplied by another rectangular matrix A, provided that the number of rows in B were equal to the number of columns in A. If A and B are both partitioned into submatrices such that the grouping of columns in A agrees exactly with the grouping of rows in B, it is not difficult to show that the product AB can equally well be obtained by treating the submatrices as elements and proceeding according to rule.

The case of square matrices of the same order, similarly and *symmetrically* partitioned, is important. Let A and B be two such matrices, and let A_{ij} henceforth denote the (ij)th submatrix in the partitioned form of A. (There will be little further occasion in this book to refer to determinantal co-factors, and the notation A_{ij} is well suited to the new concept.) Then if p, r are the orders of A_{ik}, those of B_{ki} are r, p, and those of another minor with the same k, as B_{kj}, will be r, q with the same r. For each value of k the product $A_{ik}B_{kj}$ is thus a submatrix of orders p, q; the sum $\sum_k A_{ik}B_{kj}$ can therefore be formed, and gives the (ij)th submatrix of the product AB, where the latter is in partitioned form similar to A and B. We have then, for matrices $A = [a_{ij}]$, $B = [b_{ij}]$, similarly and symmetrically partitioned,

$$AB = [\sum_k A_{ik} B_{kj}] = [C_{ij}], \quad . \quad . \quad . \quad . \quad . \quad (15)$$

where, of course, in each term of C_{ij} the order A, B is preserved.

Similarly but *unsymmetrically* partitioned square matrices A and B cannot be multiplied together by a rule of this kind; each can, however, be multiplied by the *transposed* matrix of the other, for then the partitioning of the column-groups of the multiplier agrees with that of the row-groups of the multiplicand. The transposed matrix A' is readily seen to be $A' = [A'_{ji}]$, where the minor matrix A'_{ji} is itself transposed, and for matrices A, B similarly but unsymmetrically partitioned,

$$AB' = [\sum_k A_{ik} B_{jk}]. \quad . \quad . \quad . \quad . \quad . \quad . \quad . \quad (16)$$

EXAMPLES

1. If A and B are similarly but unsymmetrically partitioned, with μ partitions into row-groups and ν into column-groups, show that the product AB' is symmetrically partitioned according to the μ row-partitions of A; and that $B'A$ is symmetrically partitioned according to the ν column-groups of A.

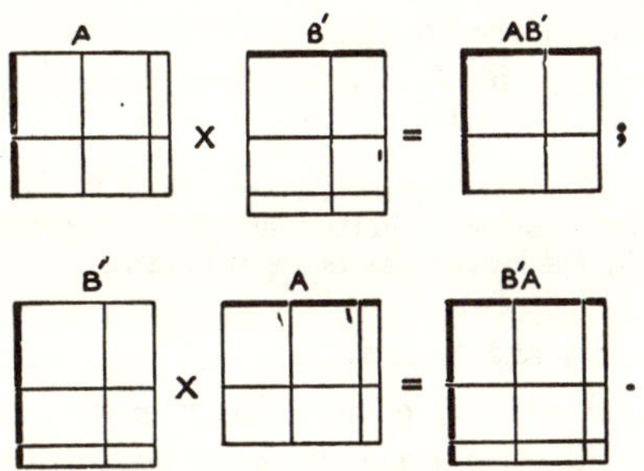

2. Distinguish by examples between a symmetrical matrix and a symmetrically partitioned matrix.

3. If
$$C = \begin{bmatrix} A & x \\ u & . \end{bmatrix} = \begin{bmatrix} a_{11} & a_{12} & a_{13} & x_1 \\ a_{21} & a_{22} & a_{23} & x_2 \\ a_{31} & a_{32} & a_{33} & x_3 \\ u_1 & u_2 & u_3 & . \end{bmatrix},$$
prove that $C^2 = \begin{bmatrix} A^2 + xu & Ax \\ uA & ux \end{bmatrix}$.

[Note that xu is a square matrix of order three, while ux is a scalar of the first order.]

4. If
$$A = \begin{bmatrix} \lambda & 1 & . & . & . \\ . & \lambda & . & . & . \\ . & . & \mu & 1 & . \\ . & . & . & \mu & . \\ . & . & . & . & \nu \end{bmatrix} = \begin{bmatrix} L & . & . \\ . & M & . \\ . & . & N \end{bmatrix},$$

where $L = \begin{bmatrix} \lambda & 1 \\ . & \lambda \end{bmatrix}$, $M = \begin{bmatrix} \mu & 1 \\ . & \mu \end{bmatrix}$, $N = \nu$, $(\lambda, \mu, \nu \neq 0)$, find the values of A^2, A^p, A^{-1}, and of any rational function $f(A)$ in terms of L, M, N.

[In general
$$f(A) = \begin{bmatrix} f(L) & . & . \\ . & f(M) & . \\ . & . & f(N) \end{bmatrix}.]$$

5. If, in the last example, L, M, N denote arbitrary square minor matrices, so that A is symmetrically partitioned, having zero minors everywhere except on the diagonal, show that $f(A)$ takes the same form.

Generalize the result, noting that the orders of the diagonal minor matrices need not necessarily be equal.

6. If
$$A = \begin{bmatrix} a & b & p & q \\ c & d & r & s \\ . & . & 1 & . \\ . & . & . & 1 \end{bmatrix} = \begin{bmatrix} B & P \\ . & I \end{bmatrix},$$

where
$$B = \begin{bmatrix} a & b \\ c & d \end{bmatrix}, \quad P = \begin{bmatrix} p & q \\ r & s \end{bmatrix}, \quad I = \begin{bmatrix} 1 & . \\ . & 1 \end{bmatrix},$$

prove that $A^n = \begin{bmatrix} B^n & \dfrac{B^n - I}{B - I} P \\ . & I \end{bmatrix}$, and that $f(A) = \begin{bmatrix} f(B) & \dfrac{f(B) - I}{B - I} P \\ . & f(I) \end{bmatrix}$.

7. The last result holds for a general square minor B, a rectangular minor P of equal depth, and unit minor I. [In the result the I occurring in the fraction has the same order as B; the lower I has its original order.]

7. Isolated Elements and Minors.

The non-zero element a_{ij} of an ordinary matrix A will be said to be *isolated* when all other elements in its own row and column are zero. It will be termed *semi-isolated* when all other elements in either its own row or its own column are zero. These terms will also be used, with the necessary changes, in speaking of submatrices A_{ij} in a partitioned matrix. The problem of reducing matrices to canonical form is largely that of carrying out isolation and semi-isolation to its utmost extent.

Thus, in Example 4 above, the three minor matrices are isolated, and so is the element ν, while the elements λ, μ are semi-isolated. The n elements of a *diagonal* matrix $[a_{ij}\delta_{ij}]$ (i.e. one in which $a_{ij} = 0$, if $i \neq j$) are isolated: and evidently n is the maximum number of elements that can be isolated in an n-rowed matrix. We shall see that any matrix can be transformed into a matrix admitting a definite number m of isolated minors, where $m \leqslant n$.

One or two elementary examples may serve to prepare the way for the use of isolated minors in dealing later with canonical forms.

First, if two square matrices P and A of the same order have similar and similarly isolated (or semi-isolated) diagonal minors, of orders $r, s, \ldots$, so also has their product PA. For example,

if
$$P = \begin{bmatrix} P_0 & P_{01} \\ . & P_1 \end{bmatrix}, \quad A = \begin{bmatrix} A_0 & A_{01} \\ . & A_1 \end{bmatrix},$$

then
$$PA = \begin{bmatrix} P_0A_0 & P_0A_{01}+P_{01}A_1 \\ & P_1A_1 \end{bmatrix}. \quad . \quad . \quad . \quad (17)$$

Here P_0, A_0, P_0A_0 are all square and of order r, while P_1, A_1, P_1A_1 are square and of order s. If both P_{01} and A_{01} are zero, then P_0, A_0, P_0A_0 are isolated in their several matrices; and so are P_1, A_1, P_1A_1. Otherwise they are semi-isolated.

Or again, consider a product of three matrices such as the following:

$$PAQ = \begin{bmatrix} I & . \\ . & P_1 \end{bmatrix} \begin{bmatrix} A_0 & A_{01} \\ . & A_1 \end{bmatrix} \begin{bmatrix} I & . \\ . & Q_1 \end{bmatrix} = \begin{bmatrix} A_0 & A_{01}Q_1 \\ . & P_1A_1Q_1 \end{bmatrix}.$$

For example,

$$\begin{bmatrix} 1 & . & . \\ . & 1 & . \\ . & -2 & 1 \end{bmatrix} \begin{bmatrix} 3 & a & b \\ . & 2 & 4 \\ . & 4 & 9 \end{bmatrix} \begin{bmatrix} 1 & . & . \\ . & 1 & -2 \\ . & . & 1 \end{bmatrix} = \begin{bmatrix} 3 & a & b-2a \\ . & 2 & . \\ . & . & 1 \end{bmatrix}.$$

In this if $A_{01} = [a, b] \neq 0$, the diagonal elements of PAQ are semi-isolated; while if a, b both vanish the isolation is complete. The point of the example lies in the fact that the minor $P_1A_1Q_1$ is obtained from the corresponding minors P_1, A_1, Q_1 exactly as the whole matrix is from PAQ. Thus in a general case further isolation might perhaps proceed within A_1, and lead step by step to the extreme stage of isolation possible in transformations of A.

8. **Historical Note.**—Matrices, considered as arrays of coefficients in homogeneous linear transformations, were of course tacitly in existence long before Cayley in 1857 proposed to develop their properties as a pure algebra of multiple number. But the intrinsic properties of the arrays were not studied for their own sake; only as much information was extracted in passing as would be useful for the application in hand, such as determinants, co-ordinate geometry, and the like. Rectangular arrays, too, had been well known from the time of the Binet-Cauchy theorem (the Theorem of Corresponding Matrices, *Invariants*, p. 79), and had found applications, for example in the normal equations of Least Squares, where the determinant is the row-by-row square of an array; and they had also been used, when premultiplied by a determinant, to express a set of determinantal equalities. Hamilton's quaternions (1843) can be regarded (*Invariants*, p. 166) as matrices of special form. But Cayley may fairly be credited with founding the general theory of matrices, for the same kind of reason as leads us to credit Vandermonde with the founding of determinants. Rectangular matrices with integer elements were investigated by H. J. S. Smith in 1861.

CHAPTER II

ELEMENTARY TRANSFORMATIONS. BILINEAR AND QUADRATIC FORMS

Many of the elementary transformations of determinants which are used in their evaluation and in the solution of linear equations may be expressed in the notation of matrices.

1. The Solution of n Linear Equations in n Unknowns.

The n equations, of which the ith is

$$\sum_{j=1}^{n} a_{ij}x_j = y_i, \qquad i = 1, 2, \ldots, n, \quad . \quad . \quad . \quad . \quad (1)$$

for n variables x_j in terms of the y_i and the n coefficients a_{ij}, are conveniently written as a single matrix equation

$$Ax = y. \quad . \quad . \quad . \quad . \quad . \quad . \quad . \quad . \quad . \quad (2)$$

If we multiply both sides of this by a square matrix P of the same order as A we obtain $PAx = Py$. In particular if $|A| \neq 0$ and P is the reciprocal of A we have

$$x = A^{-1}y, \quad . \quad . \quad . \quad . \quad . \quad . \quad . \quad (3)$$

which gives the unique solution of (1) in summary fashion. Here both x and y denote single columns of n elements: by transposition they become single rows x' and y', while Ax becomes $x'A'$, so that the same equations can also be written $x'A' = y'$. In double-suffix notation the solution is

$$x_j = \sum_{i=1}^{n} a^{ij}y_i, \qquad j = 1, 2, \ldots, n. \quad . \quad . \quad . \quad (4)$$

2. Interchange of Rows and Columns in a Determinant or Matrix.

Consideration of the identity

$$\begin{bmatrix} . & 1 & . \\ 1 & . & . \\ . & . & 1 \end{bmatrix} \begin{bmatrix} a_{11} & a_{12} & a_{13} \\ a_{21} & a_{22} & a_{23} \\ a_{31} & a_{32} & a_{33} \end{bmatrix} = \begin{bmatrix} a_{21} & a_{22} & a_{23} \\ a_{11} & a_{12} & a_{13} \\ a_{31} & a_{32} & a_{33} \end{bmatrix}$$

shows that the effect of premultiplying by a unit matrix in which the first two rows have been interchanged is to interchange those rows of A. In general let $I_{(ij)}$ denote a unit matrix in which the ith and jth rows have been interchanged without disturbing the remaining $n-2$ rows. Evidently the effect of multiplying A by such a matrix will be:

$$\left.\begin{aligned} I_{(ij)}A &: \text{interchange row}_i,\ \text{row}_j. \\ AI_{(ij)} &: \text{interchange col}_i,\ \text{col}_j. \\ I_{(ij)}AI_{(ij)} &: \text{interchange both rows and columns } i,\ j. \end{aligned}\right\} \quad (5)$$

It is to be observed that $I_{(ij)}^2 = I$, since a second interchange restores the unit matrix, and that $I'_{(ij)} = I_{(ij)}$. The double interchange thus belongs to each of two important classes of transformation, HAH^{-1} and $H'AH$.

More generally the rows, or columns, of a determinant or matrix may be deranged according to a given permutation Ω of $(123 \ldots n)$, such as $\Omega = (\alpha\beta\gamma \ldots)$. The permutation Ω of the rows of A may be effected by premultiplying by I_Ω, a matrix derived from I by subjecting the rows of I to the permutation Ω. The same permutation Ω may be impressed on the columns of A by postmultiplying by I'_Ω. Since evidently $I_\Omega I'_\Omega = I$, a permutation Ω of rows and also of columns of A belongs again to the types HAH^{-1} and $H'AH$. . . . . (6)

The only matrices of the type I_Ω which are axisymmetric are those for which Ω is a *self-conjugate* permutation. (Cf. Muir, *History of Determinants*, I, p. 60, and *Invariants*, p. 29.) Of these again only two are persymmetric, one being I and the other an important matrix with units in the *secondary diagonal*, namely, e.g.

$$J = \begin{bmatrix} \cdot & \cdot & \cdot & 1 \\ \cdot & \cdot & 1 & \cdot \\ \cdot & 1 & \cdot & \cdot \\ 1 & \cdot & \cdot & \cdot \end{bmatrix}. \quad \ldots\ldots\ldots \quad (7)$$

Evidently $J = J'$, $J^2 = I$. The effect of JA is to reverse the rows of A, while that of AJ is to reverse the columns; so that JAJ, or JAJ^{-1}, or $J'AJ$, completely reverses the matrix A.

The operation of interchanging pairs of rows or columns of A will be referred to in later sections as an operation of Type I.

3. Linear Combination of Rows or Columns in a Determinant or Matrix.

Consider again the identity

$$\begin{bmatrix} 1 & \cdot & \cdot \\ \cdot & 1 & h \\ \cdot & \cdot & 1 \end{bmatrix} \begin{bmatrix} a_{11} & a_{12} & a_{13} \\ a_{21} & a_{22} & a_{23} \\ a_{31} & a_{32} & a_{33} \end{bmatrix} = \begin{bmatrix} a_{11} & a_{12} & a_{13} \\ a_{21} + ha_{31} & a_{22} + ha_{32} & a_{23} + ha_{33} \\ a_{31} & a_{32} & a_{33} \end{bmatrix}.$$

Premultiplication of A by a unit matrix supplemented by an element h in the indicated position has effected the operation $\text{row}_2 + h\ \text{row}_3$. More generally let

$$H = I + (h)_{ij} \qquad (i \neq j)$$

denote the unit matrix with an element h inserted in the (ij)th position; then we have readily

$$\left.\begin{aligned} HA &: \ \text{row}_i + h\ \text{row}_j, \\ AH &: \ \text{col}_j + h\ \text{col}_i. \end{aligned}\right] \quad \ldots\ldots\ldots \quad (8)$$

To these may be added the following useful compositions involv ing H' and H^{-1}, which are readily seen to satisfy the relations

$$H' = I + (h)_{ji}; \quad H^{-1} = I - (h)_{ij}; \quad |H| = 1:$$

namely,

$$\left.\begin{aligned} H'AH &: \ \text{row}_j + h\ \text{row}_i,\ \text{col}_j + h\ \text{col}_i, \\ HAH^{-1} &: \ \text{row}_i + h\ \text{row}_j,\ \text{col}_j - h\ \text{col}_i. \end{aligned}\right] \quad \ldots \quad (9)$$

The notation is used in accordance with the convention in determinants: the row or column which is modified is written first, and the operations are performed in succession. (*Invariants*, p. 9.)

Operations effected by H, H', or H^{-1}, as here described, will be referred to later as of Type II.

Transformations of the type PAQ, where $|P|$ and $|Q|$ are equal to ± 1, are called *unimodular* transformations. The operations of Types I and II involve unimodular transformations.

4. Multiplication of Rows or Columns.

If in Type II the element h had been inserted in the diagonal of I, where $h \neq -1$, we should have had $H = I + (h)_{ii}$, $|H| = 1 + h$. Then, if $1 + h = k$, the rules become

$$\left.\begin{aligned} HA &: \ k\ \text{row}_i, \\ AH &: \ k\ \text{col}_i. \end{aligned}\right] \quad \ldots\ldots \quad (10)$$

For example,

$$\begin{bmatrix} 1 & . & . \\ . & k & . \\ . & . & 1 \end{bmatrix} \begin{bmatrix} a_{11} & a_{12} & a_{13} \\ a_{21} & a_{22} & a_{23} \\ a_{31} & a_{32} & a_{33} \end{bmatrix} = \begin{bmatrix} a_{11} & a_{12} & a_{13} \\ ka_{21} & ka_{22} & ka_{23} \\ a_{31} & a_{32} & a_{33} \end{bmatrix}.$$

The special use of this type of transformation, which we shall call an operation of Type III, is in an application such as the following. Let A be a diagonal matrix supplemented by non-zero elements in the superdiagonal, and let it be required to transform A in such a way as to replace those elements by units. How that may be done is visible at once, and generally, from an example such as

$$\begin{bmatrix} (ab)^{-1} & . & . \\ . & (b)^{-1} & . \\ . & . & 1 \end{bmatrix} \begin{bmatrix} \lambda & a & . \\ . & \mu & b \\ . & . & \nu \end{bmatrix} \begin{bmatrix} ab & . & . \\ . & b & . \\ . & . & 1 \end{bmatrix} = \begin{bmatrix} \lambda & 1 & . \\ . & \mu & 1 \\ . & . & \nu \end{bmatrix}. \quad (11)$$

The transformation, though not unimodular, is again of the important type HAH^{-1}, with $|H| \neq 0$.

The various transformations described in §§ 2, 3, and 4 are all of the form

$$PAQ = B, \text{ where } |P| \neq 0, |Q| \neq 0,$$

P or Q being in several cases the unit matrix I; and of course we must also have

$$|P|\ |A|\ |Q| = |B|.$$

Thus the elementary rules for simplifying determinants (or for solving linear simultaneous equations by eliminations) may be regarded as examples of matrix multiplication.

5. **Linear Transformation of Variables.**

Let the n variables x_i, which occur in (2), be subjected to a linear transformation

$$x_i = \sum_{j=1}^{n} q_{ij}\,\xi_j, \qquad i = 1, 2, \ldots, n,$$

where $Q = [q_{ij}]$ is non-singular, so that $|Q| \neq 0$. In matrix notation we have

$$Ax = y, \quad x = Q\xi, \quad AQ\xi = y. \quad . \quad . \quad . \quad . \quad (12)$$

In other words the effect of transforming the variables is equivalent to postmultiplication of A by Q, while the effect of rearranging equations, or combining them in the elementary familiar way, is equivalent to premultiplication by matrices of the type P. In canonical reduction we shall have recourse to both methods of transformation, sometimes operating on the matrix itself, sometimes setting up new systems of variables. For example, if

$$A=\begin{bmatrix}\lambda & 1 & . \\ . & \lambda & 1 \\ . & . & \lambda\end{bmatrix},\qquad B=\begin{bmatrix} . & 1 & . \\ . & . & 1 \\ \lambda^3 & -3\lambda^2 & 3\lambda\end{bmatrix}, \quad . \quad (13)$$

we may prove in the following manner that there exists a matrix H such that $HAH^{-1}=B$. As on p. 10, let there be a system of equations

$$\left.\begin{aligned}\lambda x_1 + x_2 \qquad &= y_1,\\ \lambda x_2 + x_3 &= y_2,\\ \lambda x_3 &= y_3.\end{aligned}\right] \quad . \; . \; . \; . \; . \quad (14)$$

If now x and y be transformed into ξ and η by one and the same transformation H, given by

$$\left.\begin{aligned} x_1 \qquad\qquad &= \xi_1, & y_1 \qquad\qquad &= \eta_1,\\ \lambda x_1 + x_2 \qquad &= \xi_2, & \lambda y_1 + y_2 \qquad &= \eta_2,\\ \lambda^2 x_1 + 2\lambda x_2 + x_3 &= \xi_3, & \lambda^2 y_1 + 2\lambda y_2 + y_3 &= \eta_3,\end{aligned}\right]. \quad (15)$$

we have by simple substitution

$$\left.\begin{aligned}\xi_2 \qquad &= \eta_1,\\ \xi_3 &= \eta_2,\\ \lambda^3\xi_1 - 3\lambda^2\xi_2 + 3\lambda\xi_3 &= \eta_3.\end{aligned}\right] \quad . \; . \; . \; . \quad (16)$$

The matrix of the transformation last written is B, and the whole procedure, expressed in matrix notation, is as follows:

$$Ax=y;\;\; Hx=\xi,\; Hy=\eta,\; |H|=1;\;\; B\xi=\eta=HAx=HAH^{-1}\xi. \quad (17)$$

This slight example, which shows how a matrix transformation may be carried out through the introduction of suitable new variables, and incidentally how the type of transformation HAH^{-1} may arise, epitomizes some of the work of later chapters.

6. Bilinear and Quadratic Forms.

The close relationship of a matrix A to n linear functions of n variables has just been exhibited. The n linear functions may be regarded as partial derivatives of a single function of n^2 terms, a *bilinear* form in two sets of n variables, to which A is related just as closely. If a column vector y defined by $y = Ax$ be premultiplied by a row vector u, we obtain a scalar expression

$$f = uy = uAx = \sum_{i,\ j=1}^{n}\sum^{n} u_i a_{ij} x_j. \quad . \quad . \quad . \quad . \quad (18)$$

This expression, of n^2 terms, homogeneous and linear in each of the two sets of variables u and x, is called a *bilinear form*, of matrix A. It is occasionally written $A(u, x)$. Since $\sum_{i,\ j=1}^{n}\sum^{n} x_i a_{ji} u_j$ denotes exactly the same double series of n^2 terms, we have an alternative expression for the same bilinear form, with transposed matrix:

$$f = y'u' = x'A'u'. \quad . \quad . \quad . \quad . \quad . \quad . \quad . \quad (19)$$

Bilinear forms can also arise from rectangular $m \times n$ matrices. Thus if u has m elements while x has n elements, it is easily seen that uAx is then a bilinear form of mn terms.

The matrix of a bilinear form may of course be a product PAQ, in which case we must observe that

$$vPAQ\xi = \xi'Q'A'P'v'. \quad . \quad . \quad . \quad . \quad . \quad . \quad (20)$$

The inner product uy referred to in Chapter I is evidently the special type of bilinear form which has for its matrix I, the unit matrix.

Example. $n = 2$.

$$[u_1, u_2]\begin{bmatrix} 2 & 1 \\ \cdot & \cdot \end{bmatrix}\begin{bmatrix} x_1 \\ x_2 \end{bmatrix} = 2u_1x_1 + u_1x_2 = [x_1, x_2]\begin{bmatrix} 2 & \cdot \\ 1 & \cdot \end{bmatrix}\begin{bmatrix} u_1 \\ u_2 \end{bmatrix}.$$

We may distinguish here between *matrix factors* and *scalar factors* of a form f. The matrix factors of the form uAx are u, A, x, or again x', A', u'. There may also exist scalar factors, such as $u_1(2x_1 + x_2)$ in the example above, but this is exceptional. Had we taken the matrix to be $\begin{bmatrix} 2 & \cdot \\ \cdot & 1 \end{bmatrix}$ instead of $\begin{bmatrix} 2 & 1 \\ \cdot & \cdot \end{bmatrix}$, scalar factorization would not have been possible.

If the two orderings $x'A'u'$, uAx of the form f are identical factor by factor, then f becomes a *quadratic form* of order n,

$$q = \sum_{i,j=1}^{n} a_{ij} x_i x_j, \qquad a_{ij} = a_{ji}. \quad . \quad . \quad . \quad . \quad (21)$$

The two sets of variables coincide, and the matrix A is symmetrical. Thus

$$q = x'Ax = x'A'x, \qquad A = A'. \quad . \quad . \quad . \quad (22)$$

In f the coefficient of $u_i x_j$ is a_{ij}, that of $u_j x_i$ is a_{ji}; in q these are equal. Hence the coefficient of x_i^2 is a_{ii} and that of $x_i x_j$, for $i \neq j$, is $2a_{ij}$.

7. The Highest Common Factor of Two Polynomials.

A further striking paraphrase in matrix notation of familiar ideas and processes is provided by the steps used in finding the highest common factor (H.C.F.), whether of two integers in arithmetic, or of two polynomials $\phi(\lambda)$ and $\psi(\lambda)$ in algebra. For example, consider the case of $\lambda^3 + 2\lambda^2 + 2\lambda + 1$ and $3\lambda^3 + 8\lambda^2 + 9\lambda + 4$. If we arrange the work in the usual two columns, omitting all but the successive remainders, we obtain

$$\begin{array}{r|r|r|l} \tfrac{1}{2}\lambda & \lambda^3 + 2\lambda^2 + 2\lambda + 1 & 3\lambda^3 + 8\lambda^2 + 9\lambda + 4 & 3 \\ -\tfrac{1}{6}\lambda - \tfrac{1}{3} & \tfrac{1}{2}\lambda^2 + \tfrac{3}{2}\lambda + 1 & 2\lambda^2 + 3\lambda + 1 & 4 \\ & & -3\lambda - 3 & \end{array}$$

which leads to the H.C.F. $\lambda + 1$.

Now these steps are nothing else than a *series of postmultiplications affecting a two-column matrix*

$$A = \begin{bmatrix} \phi(\lambda) & \psi(\lambda) \\ \cdot & \cdot \end{bmatrix} = \begin{bmatrix} \lambda^3 + 2\lambda^2 + 2\lambda + 1 & 3\lambda^3 + 8\lambda^2 + 9\lambda + 4 \\ \cdot & \cdot \end{bmatrix};$$

namely,

$$\text{col}_2 - 3\,\text{col}_1, \quad \text{col}_1 - \tfrac{1}{2}\lambda\,\text{col}_2, \quad \text{col}_2 - 4\,\text{col}_1, \quad \text{col}_1 + \tfrac{1}{6}(\lambda + 2)\,\text{col}_2.$$

If we denote these four steps, which are of Type II, by matrix postmultiplication, with non-singular matrices Q_1, Q_2, Q_3, Q_4, the whole process is exhibited by the relation

$$AQ_1Q_2Q_3Q_4 = \begin{bmatrix} \cdot & -3\lambda - 3 \\ \cdot & \cdot \end{bmatrix}.$$

A fifth step, of Type III, removes the factor -3, by affixing a further postmultiplier $Q_5 = \begin{bmatrix} 1 & \cdot \\ \cdot & -\tfrac{1}{3} \end{bmatrix}$: and a sixth step, of Type I, with $Q_6 = \begin{bmatrix} \cdot & 1 \\ 1 & \cdot \end{bmatrix}$, yields the result

$$AQ_1Q_2Q_3Q_4Q_5Q_6 = \begin{bmatrix} \lambda + 1 & \cdot \\ \cdot & \cdot \end{bmatrix}. \quad . \quad . \quad . \quad (23)$$

Since each element in the matrices Q_i is either zero, or a constant, or a polynomial in λ, the continued product

$$Q = Q_1Q_2Q_3Q_4Q_5Q_6 \quad . \quad . \quad . \quad . \quad . \quad . \quad (24)$$

is also a matrix having for elements polynomials in λ.

We may draw two or three important conclusions. First, that a matrix A, of the type considered here, satisfies an identity

$$AQ = \begin{bmatrix} \phi(\lambda) & \psi(\lambda) \\ \cdot & \cdot \end{bmatrix} \begin{bmatrix} q_1(\lambda) & q_3(\lambda) \\ q_2(\lambda) & q_4(\lambda) \end{bmatrix} = \begin{bmatrix} g(\lambda) & \cdot \\ \cdot & \cdot \end{bmatrix}, \quad (25)$$

where $g(\lambda)$ is the algebraic H.C.F. of the polynomials ϕ and ψ. Secondly, that the coefficient of the highest power of λ which occurs in $g(\lambda)$ can be taken to be unity. Thirdly, on performing the multiplication AQ, that the matrix Q, which is non-singular, is subject to the fundamental identity

$$\phi(\lambda)q_1(\lambda) + \psi(\lambda)q_2(\lambda) = g(\lambda) \quad . \quad . \quad . \quad . \quad (26)$$

among the polynomials concerned, and—what proves to be of less importance—to the relation $\phi(\lambda)q_3(\lambda) + \psi(\lambda)q_4(\lambda) = 0$. Lastly, that after transposition there exists an analogous relation

$$PA = \begin{bmatrix} q_1 & q_2 \\ q_3 & q_4 \end{bmatrix} \begin{bmatrix} \phi & \cdot \\ \psi & \cdot \end{bmatrix} = \begin{bmatrix} g & \cdot \\ \cdot & \cdot \end{bmatrix}. \quad . \quad . \quad (27)$$

We have purposely introduced fractional coefficients into the above example to illustrate the fact that the process is *rational*, but *not necessarily integral* in the original coefficients of ϕ and ψ. It is both rational and integral in λ. We could equally well have treated A as a vector $[\phi, \psi]$ by suppressing its second row, which is null: with the same square matrix Q we should then have $[\phi, \psi]Q = [g, 0]$.

EXAMPLES

1. The quadratic form associated with the unit matrix is

$$x'x = x_1^2 + x_2^2 + \dots + x_n^2.$$

2. Write down the quadratic and also the bilinear form associated with the diagonal matrix

$$\Lambda = \begin{bmatrix} \lambda_1 & \cdot & \cdot \\ \cdot & \lambda_2 & \cdot \\ \cdot & \cdot & \lambda_3 \end{bmatrix}.$$

$$[x'\Lambda x = \lambda_1 x_1^2 + \lambda_2 x_2^2 + \lambda_3 x_3^2; \quad u\Lambda x = \sum_{i=1}^{3} \lambda_i u_i x_i.]$$

3. To what bilinear form does a matrix A of order $m \times n$ belong?

$$[f = uAx = x'A'u' = \sum_{i=1}^{m} \sum_{j=1}^{n} u_i a_{ij} x_j.]$$

4. What is the quadratic form associated with the matrix J?

5. The elements a_{ij} of the matrix A of a quadratic form q are $\frac{1}{2}\frac{\partial^2 q}{\partial x_i \partial x_j}$.

[The determinant $\left|\frac{\partial^2 f}{\partial x_i \partial x_j}\right|$ of a form f in n variables is called (*Invariants*, p. 222) the *Hessian* of the form.]

6. If A is the matrix of a quadratic form, $H'AH$ is the matrix of another quadratic form.

7. If A is the non-singular matrix of a quadratic form, whose reciprocal is A^{-1}, prove that the quadratic form can be expressed as a determinant,

$$q = -|A| \begin{vmatrix} A^{-1} & x \\ x' & 0 \end{vmatrix},$$

where x denotes the vector matrix of the variables. (Cf. *Invariants*, p. 103.)

8. $$[112,\ 63]\begin{bmatrix} 1 & \cdot \\ -1 & 1 \end{bmatrix}\begin{bmatrix} 1 & -1 \\ \cdot & 1 \end{bmatrix}\begin{bmatrix} 1 & \cdot \\ -3 & 1 \end{bmatrix}\begin{bmatrix} 1 & -2 \\ \cdot & 1 \end{bmatrix} = [7,\ 0].$$

8. **Historical Note.**—The process of successive residuation which leads to the H.C.F. of two numbers ϕ and ψ is given by Euclid in *Book* 7, Propositions 1–3 of the *Elements*. In Proposition 33 he finds the H.C.F. of " as many numbers as we please ". In our next chapter we shall see how H. J. S. Smith made use of this process, by arranging the elements ϕ, ψ, . . . , whose H.C.F. is to be found, in rows and columns. As Smith points out,[1] his innovation was the introduction of the premultiplier P, to reinforce the effect of the postmultiplier Q in modifying a matrix A, in the form PAQ. He credits Gauss with the Q factor. (Cf. *Disq. Arith.*, §§ 213, 214, where the binary case of two rows and columns is considered.) To Hermite is attributed the introduction of a matrix of Type II in the rôle of a factor Q modifying the columns of A.

The unimodular case (p. 12) is clearly of importance when all the elements of P (or Q) are integers, in that this property is shared by the reciprocal P^{-1} (or Q^{-1}). Some writers confine the term unimodular to the case when the determinant of the matrix is unity only.

[1] *Phil. Trans.*, **151** (1861), 293–326. *Collected Works*, **1**, 367–406.

CHAPTER III

THE CANONICAL REDUCTION OF EQUIVALENT MATRICES

1. General Linear Transformation.

Suppose that the variables u and x of a bilinear form

$$f = \sum_{i=1}^{m} \sum_{j=1}^{n} u_i a_{ij} x_j = uAx \quad . \quad . \quad . \quad . \quad . \quad (1)$$

are subjected to separate non-singular linear transformations

$$u_j = \sum_{i=1}^{m} v_i p_{ij}, \quad x_i = \sum_{=1}^{n} q_{ij} \xi_j, \quad | p_{ij} | \neq 0, \, | q_{ij} | \neq 0. \quad . \quad (2)$$

When these values of u and x are substituted in f the form becomes a function of the v_i and ξ_j which is again bilinear. In matrix notation we have

$$u = vP, \quad x = Q\xi, \quad | P | \neq 0, \, | Q | \neq 0;$$

$$f = uAx = vPAQ\xi = vB\xi, \qquad B = PAQ. \quad . \quad . \quad (3)$$

Thus f has been transformed into a bilinear form

$$f_1 = \Sigma v_i b_{ij} \xi_j, \quad . \quad . \quad . \quad . \quad . \quad . \quad . \quad . \quad (4)$$

the matrix of which is derived from the original A by multiplication on either side by P and Q. The value of b_{ij} is $\sum_{\alpha=1}^{m} \sum_{\beta=1}^{n} p_{i\alpha} a_{\alpha\beta} q_{\beta j}$.

Matrices of the type PAQ constitute an important class, which will now be considered.

2. Equivalent Matrices in a Field.

Definition.—*The matrix* A *is equivalent in a field* $\mathcal{F}$ *to the matrix* B *if non-singular matrices* P *and* Q *exist such that* B = PAQ, *where the elements of* P, A, Q, *and therefore of* B, *all belong to the field.*

(i) The property of equivalence is *symmetrical*: in other words, by taking P and Q each to be the unit matrix I, we infer that A is equivalent to itself. (ii) The property is *reciprocal*: from $B = PAQ$

we derive $A = RBS$, where $R = P^{-1}$, $S = Q^{-1}$, and both R and S are non-singular. (iii) The property is *transitive*: namely if A is equivalent to B, and B to C, then A is equivalent to C. For if $B = PAQ$, $C = RBS$, then $C = RPAQS = P_1AQ_1$, where $P_1 = RP$, $Q_1 = QS$; and P, Q, R, S, P_1, Q_1 are all non-singular.

If to the above criteria, the symmetrical, the reciprocal, and the transitive, we add—what has been tacit throughout—that matrix multiplication is *associative*, we have the necessary grounds for affirming that equivalent matrices, in a given field, form a group.

The concept of a number *field*, insisted on in the definition of equivalence, is latent in all algebraical discussion. By a field $\mathcal{F}$ we mean a class of two or more numbers such that if p and q are any members whatever, equal or unequal, then $p + q$, $p - q$, $p \times q$, $p \div q$ $(q \neq 0)$ are also members of $\mathcal{F}$. To prescribe the field is to prescribe some class of numbers which secures the above condition. For example, a field $\mathcal{F}$ may consist of the rational numbers, or again of the real numbers, or again of the complex rational numbers, or again of the complex numbers. For our immediate purpose the condition means that we exclude root extractions, infinite operations and the like, and confine ourselves to addition, subtraction, multiplication, and division (except by zero), carried out a finite number of times upon the elements. The matrix A will be said to belong to a field $\mathcal{F}$ if, and only if, all its elements belong to $\mathcal{F}$.

Another important concept is that of a ring. A ring $\boldsymbol{R}$ is an assemblage of elements $p, q, \ldots$, such that $p + q$, $p - q$, pq belong to $\boldsymbol{R}$, multiplication of elements being associative and distributive. If multiplication is also commutative, we speak of a *commutative* ring. For example, the set of positive and negative integers and zero forms a commutative ring; it is not a field. If the ring possesses a unit element, which we may denote by 1, and if each element a possesses a reciprocal element b such that $ab = ba = 1$, the ring is called a *division ring*. A field is therefore a *commutative division ring*.

A common type of commutative ring in all algebraic discussion is the ring of polynomials in an arbitrary variable λ, the coefficients belonging to some prescribed field $\mathcal{F}$. For example, polynomials with rational coefficients, or with real coefficients, or with complex coefficients, form a ring. The operations of addition, subtraction and multiplication upon polynomials in the ring produce further polynomials belonging to the ring. The H.C.F. process, based as it is on products and remainders, is essentially a process within a ring. This is why the process for polynomials resembles that for integers.

3. The Equivalence of Matrices with Integer Elements.

A class of matrices of historical interest is that in which each element belongs to the ring of positive or negative integers and zero. It was shown by H. J. S. Smith that such a matrix A could be reduced by unimodular matrices P and Q to the diagonal matrix

$$PAQ = \begin{bmatrix} a_1 & \cdot & \cdots & \cdot & \cdots & \cdot \\ \cdot & a_2 & \cdots & \cdot & \cdots & \cdot \\ \cdot & \cdot & \cdot & \cdot & \cdot & \cdot \\ \cdot & \cdot & \cdots & a_r & \cdots & \cdot \\ \cdot & \cdot & \cdot & \cdot & \cdot & \cdot \\ \cdot & \cdot & \cdots & \cdot & \cdots & \cdot \end{bmatrix}, \; 0 < r \leqslant n, \quad . \quad (5)$$

where the elements of P and Q also belong to the ring. The set of r non-zero elements a_i in the diagonal uniquely characterizes A and all matrices HAK equivalent to A within the ring of operations. It was also proved by Smith that, provided the set a_i were properly arranged, each a_i contained all its predecessors as factors, thus:

$$a_1 = \beta_1, \; a_2 = \beta_1\beta_2, \; \ldots, \; a_r = \beta_1\beta_2 \ldots \beta_r. \quad . \quad (6)$$

4. Polynomials with Matrix Coefficients: λ-Matrices.

The theorem alluded to above was originally proved for matrices with integer elements. It is, however, essentially related to the problem of investigating the H.C.F. of the elements, and more generally of the minor determinants of a given order, of a matrix A; and we shall establish it for the case where the elements of A are polynomials in an arbitrary variable λ, with coefficients belonging to a prescribed field $\mathcal{F}$. Matrices of this kind have already been considered on p. 16 of Chapter II; they are called *λ-matrices*, and may be regarded as polynomials in a scalar variable but with matrix coefficients. For example

$$\begin{bmatrix} 1 & 1 \\ \cdot & 2 \end{bmatrix} + \lambda \begin{bmatrix} 2 & 1 \\ 1 & \cdot \end{bmatrix} + \lambda^2 \begin{bmatrix} 1 & 3 \\ \cdot & \cdot \end{bmatrix} = \begin{bmatrix} 1+2\lambda+\lambda^2 & 1+\lambda+3\lambda^2 \\ \lambda & 2 \end{bmatrix},$$

and in general we may have a λ-matrix

$$A(\lambda) = A_0\lambda^p + A_1\lambda^{p-1} + \ldots + A_p. \quad . \quad . \quad . \quad (7)$$

It is sometimes advantageous to introduce another variable μ in order to make the expressions homogeneous, when the λ-matrix $A(\lambda)$ is replaced by a binary p-ic with matrix coefficients or, if we please, by a single matrix $A(\lambda, \mu)$ of which each element is a scalar binary p-ic.

The notation of congruence with respect to a modulus can also be usefully applied to matrices. Thus the statement $B \equiv C \bmod \lambda$ will be taken to mean that each element of the matrix $B - C$ is zero or else contains λ as a factor. The notation will be understood at once from the example

$$\begin{bmatrix} 1+2\lambda+\lambda^2 & 1+\lambda+3\lambda^2 \\ \lambda & 2 \end{bmatrix} \equiv \begin{bmatrix} 1 & 1 \\ \cdot & 2 \end{bmatrix} \bmod \lambda. \qquad (8)$$

5. **The H.C.F. Process for Polynomials.**

An aspect of the H.C.F. process, from the standpoint of matrices, has been considered on p. 16. We take the opportunity of recording here more fully for future reference the essential algebraic features of the process.

We begin with two polynomials $\phi(\lambda)$ and $\psi(\lambda)$, of degrees m and n, where $m \leqslant n$, with scalar coefficients belonging to a field $\mathcal{F}$, and such that $\phi(\lambda)$ has unity for the coefficient of its highest power:

$$\phi(\lambda) = \lambda^m - c_1\lambda^{m-1} - c_2\lambda^{m-2} - \ldots - c_m.$$

By ordinary long division of ψ by ϕ a polynomial quotient and remainder are uniquely obtained: and since the process is rational and terminating, all coefficients in the quotient and remainder belong also to $\mathcal{F}$. We have then

$$\psi(\lambda) = q(\lambda)\phi(\lambda) + r(\lambda), \quad \ldots\ldots \quad (9)$$

where the quotient $q(\lambda)$ is of degree $n - m$, and the remainder $r(\lambda)$ of degree necessarily less than m. If ϕ is a factor of ψ then r is zero.

Again, by continuing according to Euclid's algorithm, the ratio $\psi(\lambda)/\phi(\lambda)$ is expressed as a continued fraction, the last convergent, $h(\lambda)/k(\lambda)$, providing the identity

$$\psi(\lambda)k(\lambda) - \phi(\lambda)h(\lambda) = g(\lambda), \quad \ldots\ldots \quad (10)$$

where $g(\lambda)$ is the algebraic H.C.F. of $\phi(\lambda)$ and $\psi(\lambda)$, as in § 7, (26), of p. 16, and all five expressions ϕ, ψ, h, k, g are polynomials with coefficients belonging to $\mathcal{F}$. It is usually convenient to divide h, k, g throughout by the coefficient of the highest power of λ in g; the coefficients of the polynomials still remain in the field $\mathcal{F}$, and the algebraic factors are unaltered. If ϕ and ψ are algebraically prime to each other their H.C.F. is then unity.

The procedure and results just given can be transferred at once to the case of polynomials in a matrix A with scalar coefficients. (These are to be distinguished from λ-matrices, which are polynomials in a *scalar* λ with *matrix* coefficients.) For a square matrix A is permutable in multiplication with powers A^p of itself, and it follows at once that scalar polynomials in A obey the commutative law of multiplication and so form a commutative ring. Hence we have results corresponding to (9) and (10) above:

$$\left.\begin{aligned}\phi(A) &= A^m - c_1 A^{m-1} - c_2 A^{m-2} - \ldots - c_m I,\\ \psi(A) &= q(A)\phi(A) + r(A),\\ \psi(A)k(A) - \phi(A)h(A) &= g(A).\end{aligned}\right\} \quad . \quad . \quad (11)$$

Each of the identities (11) can also be interpreted as n^2 scalar identities, usually of a more complicated kind, referring to the equality of pairs of corresponding elements in the matrices on either side.

Example.—If $\phi(A) = A^2 - 2A + I$, $\psi(A) = A^3 - 4A^2 + 6A - 3I$, then $\qquad g(A) = A - I,\ h(A) = A - 2I,\ k(A) = I.$

6. Smith's Canonical Form for Equivalent Matrices.

Theorem I.—*Every λ-matrix of order* n *and rank* r *can be reduced by rational transformations to the equivalent diagonal form containing exactly* r *isolated elements,*

$$D = PAQ = \begin{bmatrix} E_1(\lambda) & \cdot & \cdots & \cdot & \cdots & \cdot \\ \cdot & E_2(\lambda) & \cdots & \cdot & \cdots & \cdot \\ \cdot & \cdot & \cdot & \cdot & \cdot & \cdot \\ \cdot & \cdot & \cdots & E_r(\lambda) & \cdots & \cdot \\ \cdot & \cdot & \cdot & \cdot & \cdot & \cdot \\ \cdot & \cdot & \cdots & \cdot & \cdots & \cdot \end{bmatrix}, 0 < r \leqslant n, \quad (12)$$

where $\mathrm{E}_1(\lambda) = \delta_1$, $\mathrm{E}_2(\lambda) = \delta_1\delta_2, \ldots,$ $\mathrm{E}_r(\lambda) = \delta_1\delta_2\ldots\delta_r$, *each* δ_i *being unity or a non-zero polynomial in* λ *having unity for coefficient of its highest power and having all its coefficients in the field* $\mathcal{F}$.

The proof of this theorem will be gradational, and will consist in operating successively on A with λ-matrices P_i and Q_i, so chosen that the matrices P_1AQ_1, $P_2P_1AQ_1Q_2, \ldots$ assume more and more the diagonal aspect. The operations used will be exclusively of the Types I, II, and III of Chapter II. Part of the work is really implicit in § 7, p. 16, but we shall describe it in full.

Proof.—If A is the null matrix, it is already in canonical form with $r = 0$. If it is not, there will be at least one non-zero element of lowest degree in λ. Let this be a_{ij}; then the interchange $I_{(1i)}\,AI_{(1j)}$ will bring a_{ij} to the position a_{11}. This, the *first step*, is carried out by an operation of Type I. We now refer to the former a_{ij} as a_{11}.

Next, the elements of the first row, with the exception of a_{11}, may all be zero. If this is not so, there will be an $a_{1j} \neq 0$. Let this polynomial a_{1j} be divided by a_{11}, yielding a quotient q_{1j} and a remainder r_{1j}, which are also polynomials in λ. By the operation $\text{col}_j - q_{1j}\,\text{col}_1$ the element a_{1j} is replaced by r_{1j}, which is either zero or of lower degree than a_{11}. The operations must end, since each r_{1j} or r_{i1} that is taken as a new a_{11} is of lower degree than its predecessor. In the latter case the first step (the interchange) may be repeated, and in a finite number (less than p) of such combined steps the remainders are exhausted and a_{1j} is replaced by zero. We can apply this process in turn to all the elements of the first row. In the end they are all replaced by zero, except a_{11}. In a similar way we now clear the first column, then the first row again, and so on, until every element in the first row and first column, except a_{11}, is zero.

Thirdly, if there still remain any element a_{ij} not exactly divisible by a_{11}, we shall have $a_{11} \neq 0$, $a_{i1} = a_{1j} = 0$, $a_{ij} = q_{ij}\,a_{11} + r_{ij}$, where $r_{ij} \neq 0$. Then the operations $\text{col}_j + q_{ij}\,\text{col}_1$, $\text{row}_i - \text{row}_1$ replace a_{ij} by r_{ij}, at the expense of introducing $-a_{11}$ and $q_{ij}a_{11}$ at the places $(i, 1)$ and $(1, j)$. Had r_{ij} been zero no advantage would have been gained by this step, for the removal of the introduced $-a_{11}$ would merely have restored the former situation, causing " perpetual check ".) Now the first step applies again, the second follows and, if necessary, the third, and so on, until at last the third is found unnecessary; unnecessary because every non-zero a_{ij} will be divisible by $a_{11} = \beta_1$. At this stage A has been reduced to

$$P_1AQ_1 = \begin{bmatrix} \beta_1 & \cdot \\ \cdot & A_{11} \end{bmatrix}, \qquad \beta_1 \neq 0, \qquad A_{11} \equiv 0 \bmod \beta_1. \quad . \quad (13)$$

We can now write $A_{11} = \beta_1 B_1$, where B_1 is a λ-matrix of the $(n - 1)$th order. A similar course of operations brings B_1 to the equivalent form

$$P_2B_1Q_2 = \begin{bmatrix} \beta_2 & \cdot \\ \cdot & \beta_2 B_2 \end{bmatrix},$$

where B_2 is a λ-matrix of order $(n - 2)$. Hence

$$\begin{bmatrix} 1 & \cdot \\ \cdot & P_2 \end{bmatrix} \begin{bmatrix} \beta_1 & \cdot \\ \cdot & \beta_1 B_1 \end{bmatrix} \begin{bmatrix} 1 & \cdot \\ \cdot & Q_2 \end{bmatrix} = \begin{bmatrix} \beta_1 & \cdot & \cdot \\ \cdot & \beta_1\beta_2 & \cdot \\ \cdot & \cdot & \beta_1\beta_2 B_2 \end{bmatrix}. \quad (14)$$

So the reduction proceeds until either the nth row is reached or else after an earlier rth row the minor matrix B_r is null. Finally if b_i is the coefficient of the highest power of λ in the non-zero polynomial β_i, then $b_i \neq 0$ and we write $\beta_i = b_i \delta_i$. By operations of Type III we may divide the ith row by b_i for each i.

The reduced matrix is now in the form D; and since all the operations used have been of Types I, II, or III we have $PAQ = D$, with $|P| \neq 0$, $|Q| \neq 0$. The proof is therefore complete.

7. The H.C.F. of m-rowed Minors[1] of a λ-Matrix.

The leading element $E_1(\lambda)$ of the canonical form $D = PAQ$ is the H.C.F. of the elements of A, or of D. This is a special case of a more general theorem regarding the H.C.F. of the m-rowed minors of A, or of D.

Theorem II.—*The product* $G_m(\lambda) = E_1(\lambda)E_2(\lambda) \ldots E_m(\lambda)$ *is the H.C.F. of all the* m*-rowed minor determinants of* D *and also of* A.

Proof.—Every non-zero minor of order m in D is a product of m diagonal elements $E_{k_1}(\lambda)$, $E_{k_2}(\lambda)$, ..., $E_{k_m}(\lambda)$. The suffixes may be taken in ascending order and obviously satisfy the conditions $k_1 \geqslant 1$, $k_2 \geqslant 2$, ..., $k_m \geqslant m$. Hence $G_m(\lambda)$ is a common factor of all the m-rowed minors; and it is the highest common factor.

To prove the second part of the theorem, consider the actual relation

$$PAQ = \begin{bmatrix} E_1 & . & \cdots & . & \cdots \\ . & E_2 & \cdots & . & \cdots \\ . & . & . & . & . \\ . & . & \cdots & E_r & \cdots \\ . & . & . & . & . \\ . & . & \cdots & . & \cdots \end{bmatrix} = D.$$

Since P and Q are products of matrices of the Types I, II, and III, the determinants $|P|$ and $|Q|$ are products of non-zero constants—units in cases I and II, factors b_1 in case III. Hence $|P|$ and $|Q|$ are independent of λ. It follows at once that the reciprocal matrices P^{-1} and Q^{-1} are λ-matrices in the field, with determinants independent of λ.

Now, by the Binet-Cauchy theorem (Theorem of Corresponding Matrices) any m-rowed minor of PA is a linear function of the m-rowed minors of A. For example (cf. *Invariants*, p. 81),

[1] For the rest of this chapter "minor" will mean "minor determinant".

if
$$A = \begin{bmatrix} a_1 & b_1 & c_1 \\ a_2 & b_2 & c_2 \\ a_3 & b_3 & c_3 \end{bmatrix}, \quad P = \begin{bmatrix} p_1 & p_2 & p_3 \\ q_1 & q_2 & q_3 \\ r_1 & r_2 & r_3 \end{bmatrix},$$

then
$$PA = \begin{bmatrix} (pa) & (pb) & (pc) \\ (qa) & (qb) & (qc) \\ (ra) & (rb) & (rc) \end{bmatrix}, \qquad (pa) = \Sigma\, p_i a_i;$$

and if $(pq)_{ij}$ denotes $p_i q_j - p_j q_i$, then the two-rowed minor $\begin{vmatrix} (pa) & (pb) \\ (qa) & (qb) \end{vmatrix}$ is equal to $\sum_{i,j} (pq)_{ij}(ab)_{ij}$, for $i, j = 2, 3;\ 3, 1;\ 1, 2$: that is, it is equal to a linear function of minors $(ab)_{ij}$ of A.

For the same reason every m-rowed minor of PAQ is a linear function of m-rowed minors of A. Hence every common factor, including the H.C.F., of the m-rowed minors of A is a factor of each m-rowed minor of D; and since $A = P^{-1}DQ^{-1}$, the same is true with D and A interchanged. Hence the m-rowed minors of each have the same H.C.F., namely $G_m(\lambda)$.

8. **Equivalent λ-Matrices.**

In using the word *equivalent* in the course of the preceding demonstrations we have anticipated the formal definition of equivalent λ-matrices.

Two λ-matrices A *and* B *are said to be equivalent in the field* $\mathcal{F}$ *when λ-matrices* P *and* Q *exist in the field, such that their determinants* $|\mathrm{P}|$ *and* $|\mathrm{Q}|$ *are non-zero and independent of* λ, *and such that* $\mathrm{B} = \mathrm{PAQ}$.

The definition includes the definition of equivalence of p. 19, which follows by taking elements all of zero degree in λ. It also entails all the group properties previously enumerated: these may be verified without difficulty. Accordingly the diagonal canonical form D is equivalent to A and to all the equivalents of A: and the question arises whether the form D is unique or not. It is a remarkable fact that A and B can be rectangular, the rows conforming with the square matrix Q, and the columns with P. An illustration will occur on p. 30.

Theorem III.—*The λ-matrix* A *of order* n *and all its equivalent matrices have a unique diagonal canonical form with a diagonal of* r *non-zero elements* $\mathrm{E}_1(\lambda), \mathrm{E}_2(\lambda), \ldots, \mathrm{E}_r(\lambda)$, *followed by* n − r *zeros.*

Proof.—By Theorem II, any two equivalent matrices A and $B = PAQ$ have the same polynomial $G_m(\lambda)$ as H.C.F. of their own

m-rowed minors. By Theorem I, the matrix B has its own canonical form of diagonal elements, say $E_1'(\lambda)$, $E_2'(\lambda)$, ... It follows that

$$E_1'(\lambda)E_2'(\lambda)\ldots E_m'(\lambda)=G_m'(\lambda)=G_m(\lambda)=E_1(\lambda)E_2(\lambda)\ldots E_m(\lambda), \quad (15)$$

for $m=1, 2, \ldots, r$, and $E'_{r+i}(\lambda)=E_{r+i}(\lambda)=0$. Taking the values of m in succession we find $E_i'(\lambda)=E_i(\lambda)$, so that A and B have an identical canonical form D.

9. **Observations on the Theorems.**

We have arrived at an integer r $(0 < r \leqslant n)$ which is invariant for all equivalent transformations of a matrix A. We have also obtained a unique sequence of r non-zero polynomials $E_i(\lambda)$ each of which is a factor of all its successors and has unity for coefficient of its highest power of λ.

The number r is the *rank* (*Invariants*, p. 73) of the canonical matrix D and all its equivalents: for $G_r(\lambda)$ is a non-zero minor of the rth order in D, while every minor of higher order is zero: and this defines the rank of D. As an alternative and equivalent test, D contains exactly r rows (or columns) and no more, which are linearly independent (see Chapter V) in the field $\mathcal{F}$. Since each m-rowed minor of A is a linear combination of m-rowed minors of D, the matrix A satisfies the same condition: so that its rank is also r.

It is to be noted that the polynomials $E_i(\lambda)$ are rational, though not necessarily integral, functions of the elements of A; they are polynomial invariants for all matrices equivalent to A within the ring. They have been called the *invariant factors* of A.

As for the matrices P and Q which appear in the reduced form PAQ, we have not proved that they are unique. Evidently they are not, for if K be any non-singular diagonal matrix of the order of A, it is easy to see that $KPAQK^{-1}$ is identical with PAQ. More generally if A is of rank r, the principal submatrix of K formed by the first r rows and columns can be purely diagonal, while that formed by the last $n-r$ can be entirely arbitrary, except in so far as K is to be non-singular.

The relationship of Theorems II and III to Theorem I is best regarded from the standpoint of compound matrices. (*Invariants*, p. 87.) If the $\binom{n}{m} \times \binom{n}{m}$ minors of the mth order in A be arranged in a square of $\binom{n}{m}$ rows and columns, priority of location in these being decided by the priority of the rows and columns from which the minors are taken (the order of row-groups being thus the same as for column-groups), the matrix so derived is called the mth compound of A and may be denoted by $A^{(m)}$. The generalized multiplication theorem of determinantal arrays, found simultaneously and independently by Cauchy and Binet in 1812, is then co-extensive with the important theorem,

$$[ABC\ldots K]^{(m)} = A^{(m)}B^{(m)}\ldots K^{(m)}, \quad . \quad . \quad . \quad (16)$$

that is, *the* m*th compound of the product of a number of matrices of the same order is element for element identical with the product of the* m*th compounds of the several matrices.* When this is applied to Theorem I we have at once $P^{(m)}A^{(m)}Q^{(m)} = D^{(m)}$, where the determinants $|P^{(m)}|$ and $|Q^{(m)}|$, by Sylvester's theorem on compound determinants (*Invariants*, p. 87), are the $\binom{n-1}{m-1}$th powers of $|P|$ and $|Q|$ and are thus non-singular and independent of λ. A consideration of the form of the diagonal matrix $D^{(m)}$ makes Theorems II and III immediately apparent. Here, as in so many other contexts, it is illuminating to think of typical properties and transformations of determinants or matrices as being reflected at the same time by similar features in all their compounds. The remarkable maturity, considering the early date, of Cauchy's great memoir of 1812 is in great measure due to the fact that he envisaged with an array all its "*systemes dérivés*", that is, its compound matrices.

A matrix in $\mathcal{F}$ with constant elements a_{ij} can be reduced by operations in $\mathcal{F}$ to $PAQ = D$, where each α_r is a unit element. For we might regard this as the reduction of $A - \lambda I$ for the case $\lambda = 0$. Evidently, since $|P|$ and $|Q|$ are constants, P and Q remain non-singular when $\lambda = 0$. Thus we see that the rank r of D (possibly lower than the rank of $A - \lambda I$) is invariant, in that all matrices equivalent to A in $\mathcal{F}$ have rank r.

EXAMPLES

1. Reduce to equivalent canonical form

$$A = \begin{bmatrix} 1+(1+\lambda)(1-\lambda^2) & 1-\lambda^2 & 1-\lambda^2 \\ 1-\lambda^2 & (1-\lambda)(1-\lambda^2) & (1-\lambda)(1-\lambda^2) \\ 1+(1+\lambda)(1-\lambda^2) & 1-\lambda^2 & 2(1-\lambda^2) \end{bmatrix}.$$

[$E_1(\lambda) = 1$, $E_2(\lambda) = 1-\lambda^2$, $E_3(\lambda) = (1-\lambda^2)(1-\lambda-\lambda^2+\lambda^4)$.]

2. $A = \begin{bmatrix} 1 & 2 & 1 \\ 4 & 11 & 10 \\ 2 & 4 & 17 \end{bmatrix}$. Reduce A to canonical form $D = PAQ$, where the elements of P, Q and the diagonal matrix D are integers or zero, and $|P| = |Q| = 1$. The operations of Types I, II, III may be used, but h in Type II must be an integer, and k in Type III (p. 13) must be ± 1.

10. The Singular Case of n Linear Equations in n Variables.

The case of n linear equations in n variables with non-singular matrix has been considered on p. 10. When the matrix is singular the method there given cannot be employed; but a parametric solution can be arrived at by the aid of the canonical form $D = PAQ$.

Consider the n linear equations given by $Ax = y$, where $|A| = 0$, and suppose that an equivalent canonical form $D = PAQ$ has been found, with r non-zero consecutive diagonal elements $\alpha_1, \alpha_2, \ldots, \alpha_r$, all rational in the coefficients a_{ij}. P and Q are non-singular, and belong to the same field as the a_{ij}. Let Q^{-1} be calculated, and let n new variables ξ_i be introduced through the linear relations

$$\xi = Q^{-1}x. \quad . \quad . \quad . \quad . \quad . \quad . \quad . \quad (17)$$

Since $Ax = y$, we have $PAx = Py$ and hence

$$D\xi = PAQ\xi = PAQQ^{-1}x = PAx = Py.$$

In the equation $D\xi = Py$ the left-hand member $D\xi$ consists of a column of r non-zero elements $\alpha_1\xi_1, \alpha_2\xi_2, \ldots, \alpha_r\xi_r$ followed by $n - r$ zeros; the right-hand member Py denotes a single column matrix consisting of n homogeneous linear functions of the y_i, the coefficients in these being the elements of the respective rows of P.

The equation $D\xi = Iy$ can evidently be solved for the ξ_i *if the last* n − r *elements of the column matrix* $\eta =$ Py *are zero.* If this is so, we have $\alpha_1\xi_1 = \eta_1, \alpha_2\xi_2 = \eta_2, \ldots, \alpha_r\xi_r = \eta_r$, which give $\xi_i = \eta_i/\alpha_i$. The remaining $n - r$ of the ξ_i may be arbitrary: they do not affect the value of $D\xi$. Since $x = Q\xi$, each x_i is therefore a linear function of all the ξ_i, with coefficients in $\mathcal{F}$.

We see then that the general solution for the x_i, subject to the condition insisted upon in the preceding paragraph, involves exactly $n - r$ parameters $\xi_{r+1}, \ldots, \xi_n$: for all other expressions used in the argument have been obtained by rational steps involving only the n^2 elements a_{ij} and the n variables y_i. Since P and Q are non-singular, each step of the argument is reversible.

If any of the last $n - r$ elements η_{r+i} is non-zero no solution exists, and the original equations are said to be *incompatible.* For example

$$x_1 + x_2 = 1, \quad x_1 + x_2 = 2.$$

We may put the condition of compatibility in a more general form by considering the augmented matrix

$$C = [A, y] \quad . \quad . \quad . \quad . \quad . \quad . \quad . \quad . \quad (18)$$

of n rows and $n+1$ columns, the ith row being $a_{i1}, a_{i2}, \ldots, a_{in}, y_i$: namely

The n *non-homogeneous linear equations* Ax = y *are compatible and soluble in terms of* n − r *arbitrary parameters, if and only if the rank of* C, *the augmented matrix, is equal to* r, *that of* A.

Proof.—Consider the identity

$$P[A, y]\begin{bmatrix} Q & \cdot \\ \cdot & 1 \end{bmatrix} = [PAQ, Py], \quad . \quad . \quad . \quad . \quad (19)$$

the matrices on the left being of order $n \times n$, $n \times (n+1)$, $(n+1) \times (n+1)$ respectively. The matrix product is therefore equal to $[D, \eta]$, a matrix of order $n \times (n+1)$, the rank of which must be r at least, since one of its r-rowed minors (included in D) is the non-zero product $a_1 a_2 \ldots a_r$. Further, its rank can only exceed r if at least one element η_{r+i} $(i > 0)$ in the last $n - r$ rows and the final column is non-zero. If this is not so, then the rank is r.

But Q is non-singular; so therefore is $\begin{bmatrix} Q & \cdot \\ \cdot & 1 \end{bmatrix}$, since its determinant is equal to $|Q|$. Hence the rank of $[A, y]$, since it is unaltered by multiplication with non-singular matrices (cf. *Invariants*, p. 84), is that of $[D, \eta]$. Hence finally if, and only if, the matrices C and A have the same rank r, the last $n - r$ elements of Py are zero, and the equations $Ax = y$ are compatible and parametrically soluble.

EXAMPLES

1. If r is the rank of $\begin{bmatrix} A & B \\ C & D \end{bmatrix}$, which is partitioned into $r + s$ rows and $r + s$ columns, where A is a non-singular $r \times r$ matrix, prove that $D = CA^{-1}B$. Prove also the converse.

[Consider the identity $\begin{bmatrix} I & \cdot \\ P & I \end{bmatrix} \begin{bmatrix} A & B \\ C & D \end{bmatrix} \begin{bmatrix} I & Q \\ \cdot & I \end{bmatrix} = \begin{bmatrix} A & \cdot \\ \cdot & D - CA^{-1}B \end{bmatrix}$ where $PA + C = 0 = AQ + B$.]

2. If x and h are column vectors each of r components and y and k are such with s components, show that the equations

$$Ax + By = h, \qquad Cx + Dy = k,$$

for x and y are compatible, for the above matrix coefficients A, B, C, D, if and only if

$$CA^{-1}h + k = 0.$$

Solve the equations.

[$x = A^{-1}h - A^{-1}By$, giving the x in terms of s parameters y_i.]

11. **Historical Note.**—Sylvester in 1851 and Cayley in 1854 had classified some families of quadrics by invariant factors into all possible canonical types. The diagonal canonical form D was found for a matrix of integers by H. J. S. Smith in 1861, as already mentioned. The invariant properties and the divisibility of each E_1 by its predecessors were demonstrated, and the arbitrary nature of P and Q was explicitly stated; the last-named fact was proved in a later paper of 1873. The extension to λ-matrices was carried out by Frobenius in 1878. Weierstrass, in 1868, in considering the simultaneous transformation of families of bilinear forms, had established the invariance of a set of polynomials closely related to $E_i(\lambda)$ which he named *Elementartheiler*, a term taken over later by Frobenius in the sense of " invariant factor ", as we have defined it. The relation of the original Elementartheiler to the invariant factors $E_i(\lambda)$ will be considered in a later chapter.

An excellent account of the development of the theory of invariant factors, from the standpoint of determinants, is to be found in Muir's *History*, IV, pp. 435–453. Full references are there given.

The respective claims of Binet and Cauchy to the generalized multiplication theorem of determinants are discussed at length by Muir, *History*, I, pp. 123–130. One way of putting the matter seems to be that the equivalent of Cauchy's contribution is simply (16) above, all the matrices concerned being square, while that of Binet is (16), with matrices of order alternately $m \times n$ and $n \times m$. The extension of (16) to generally conformable matrices includes both.

The proof of (16) for the case of two matrices $[PA]^{(m)} = P^{(m)}A^{(m)}$ has virtually been given on p. 26. The equality of the ijth elements on the left and right of this identity is a direct consequence of the Binet-Cauchy Theorem (*Invariants*, p. 81). The proof for the case of three or more follows immediately from that of two matrices.

The treatment and results in the Theorems I, II, and III evidently apply, with the slightest modification, to rectangular as well as to square matrices.

CHAPTER IV

Subgroups of the Group of Equivalent Transformations

In the remaining part of this survey we shall consider only matrices having elements which are either constant numbers, real or complex, or are at the most linear in λ. At the same time we shall impose on the matrices P and Q in an equivalent transformation PAQ various restrictions, giving rise to several different subgroups, all of some importance, within the equivalent group. Two of these subgroups, characterized by $H'AH$ and HAH^{-1}, have been touched upon in earlier sections.

Certain of these transformations are intimately associated with special types of square matrix A, as for example the *orthogonal* transformation associated with *axisymmetric* matrices, the *unitary* transformation with *Hermitian* matrices, and so on. We proceed to define these special matrices and to show how inevitably they arise when matrix operations, and the types of matrix which preserve their type under these operations, are systematically classified.

1. **Matrices of Special Type, Symmetric, Orthogonal, &c.**

To the operations upon A expressed by A' and A^{-1}, as already defined by transposition and inversion, we may add another, $\bar{A}$, implying that if the elements of A are complex scalars the corresponding elements of $\bar{A}$ are their respective complex conjugates. The matrix $\bar{A}$ is naturally called the *conjugate* of A, while $\bar{A}'$, the transposed conjugate, is called the *associate* (begleitende) of A. We have then four principal operations, $-A$, A', A^{-1}, and $\bar{A}$, each of which has the reflexive property, in that when performed twice it reproduces the original matrix A:

$$-(-A) = A, \quad (A')' = A, \quad (A^{-1})^{-1} = A, \quad \bar{\bar{A}} = A. \qquad (1)$$

(In the algebra of complex *scalars* there are three operations only, $-a$, a^{-1}, and $\bar{a}$.) Moreover, as may readily be verified, any two of the four operations, say T_α and T_β, are *permutable*, in that we have $T_\alpha T_\beta \,.\, A = T_\beta T_\alpha \,.\, A$. It is well to note in passing, however, that while the reversal law holds for transposition and reciprocation of products, it is not required for conjugation:

$$(AB)' = B'A', \quad (AB)^{-1} = B^{-1}A^{-1}, \quad \text{but } (\overline{AB}) = \overline{A}\,\overline{B}. \qquad (2)$$

When due regard is paid to the permutable property, we have in all 2^4 matrices derived from A by combining the four operations in all possible ways, the sixteen related matrices being $\pm A$, $\pm A'$, $\pm A^{-1}$, $\pm \overline{A}$, $\pm(A')^{-1}$, $\pm \overline{A}'$, $\pm \overline{A}^{-1}$, $\pm(\overline{A}')^{-1}$. These are in general all distinct, but in particular cases two or more of them may be identical, and it is interesting to observe that by identifying A with various other members of the set we obtain most of the important special types of matrices. Excluding trivial cases (for example, if $A = -A$ the matrix A can only be null), we may note the following:

(i) If $A = A'$, A is *symmetric*. ($a_{ij} = a_{ji}$.)

(ii) If $A = A^{-1}$, A is *involutory*.

(iii) If $A = \overline{A}$, A is *real*. (a_{ij} is real for all i, j.)

(iv) If $A = (A')^{-1}$, A is *orthogonal*.

(v) If $A = \overline{A}'$, A is *Hermitian*. ($a_{ij} = \bar{a}_{ji}$.)

(vi) If $A = (\overline{A}')^{-1}$, A is *unitary*.

The type $A = \overline{A}^{-1}$, which does not appear in the above list, seems up to the present to have found no application and to have received no special name. When we take into account the negative operation we derive further types, of which only three are of importance:

(vii) If $A = -A'$, A is *skew symmetric* or *alternate*. ($a_{ii} = 0$, $a_{ji} = -a_{ij}$.)

(viii) If $A = -\overline{A}$, A is *pure imaginary*. (a_{ij} is pure imaginary for all i, j.)

(ix) If $A = -\overline{A}'$, A is *skew Hermitian*, or *Hermitian alternate*. ($a_{ij} = -\bar{a}_{ji}$.)

EXAMPLES

1. The matrix $\begin{bmatrix} 1 & i & 1+i \\ -i & 2 & 1 \\ 1-i & 1 & 3 \end{bmatrix}$ is Hermitian.

2. If A is any matrix, $A'A$ is symmetric, and $\bar{A}'A$ is Hermitian.

$$[(A'A)' = A'(A')' = A'A: \text{ also } \bar{A}'A = (\overline{\bar{A}'A})'.]$$

3. The matrix $Q = \frac{1}{2}(A + A')$ is symmetric, while $S = \frac{1}{2}(A - A')$ is skew symmetric. Thus any square matrix A may be resolved into the sum of a symmetric and a skew symmetric matrix, $Q + S$. In the same way any square matrix of complex elements is the sum of a Hermitian and a skew Hermitian matrix.

4. All integral powers of a symmetric matrix are symmetric. Positive odd powers of a skew symmetric matrix are skew symmetric, positive even powers are symmetric. Extend this to Hermitian matrices.

5. A skew symmetric matrix of odd order is singular. Prove this, and deduce that the reciprocal of a skew symmetric matrix of even order is skew symmetric.

6. If $A = \bar{A}^{-1}$, and $A = X + iY$, prove that $XY = YX$, and $X^2 + Y^2 = I$.

7. The algebraic properties of pairs of matrices of the second order of the type $\begin{bmatrix} a & -b \\ b & a \end{bmatrix}, \begin{bmatrix} a & b \\ -b & a \end{bmatrix}$ are in correspondence with those of the complex scalars $a + ib, a - ib$.

8. The continued product $[z, 1] \begin{bmatrix} a & b \\ c & d \end{bmatrix} \begin{bmatrix} \bar{z} \\ 1 \end{bmatrix}$ is equal to the complex scalar $az\bar{z} + bz + c\bar{z} + d$. If $z = x + iy$ $\bar{z} = x - iy$, the scalar is real if, and only if, the matrix $\begin{bmatrix} a & b \\ c & d \end{bmatrix}$ is Hermitian. [A bilinear form in conjugate complex variables, with Hermitian matrix, is real, and is called a *Hermitian form.*]

9. Prove that if A belongs to any one of the special types described above then each of the sixteen allied matrices is of the same type.

2. Axisymmetric, Hermitian, Orthogonal, and Unitary Matrices.

The real axisymmetric (briefly "symmetric") type is a special case of the Hermitian. Thus if $A = X + iY$ is Hermitian, all the elements of X and Y being real, we must have $X' + iY' = X - iY$, so that $X' = X$, $Y' = -Y$. Thus the real part of the Hermitian matrix A is symmetric, the imaginary part skew symmetric; or $A = Q + iS$, where Q is real symmetric, S is real skew symmetric. Hence if A is purely real it must be symmetric; on the other hand a complex axisymmetric matrix is not Hermitian.

In a similar manner the real orthogonal type is a special case of the unitary. For in the latter case if $A = X + iY$ then $\bar{A}' = X' - iY'$, and so we have $(X' - iY')(X + iY) = I$. If $Y = 0$ then $Y' = 0$ and $X'X = I$, a relation which defines the real orthogonal case. The complex orthogonal case is to be distinguished from the unitary.

The unitary and the orthogonal types possess *group* properties.

For if A and B are both unitary, then $\bar{A}'A = \bar{B}'B = I$, and so by the reversal and the associative laws,

$$(\overline{AB})'AB = \overline{B}'\overline{A}'AB = \overline{B}'(\overline{A}'A)B = \overline{B}'B = I. \quad . \quad (3)$$

Hence AB is also unitary. Similarly if A and B are orthogonal so is AB. That the reciprocals of these types are respectively of the same type was set for proof in Example 9 above.

EXAMPLES

1. The matrix $\begin{bmatrix} \cos\theta & \sin\theta \\ -\sin\theta & \cos\theta \end{bmatrix}$ is orthogonal.

2. The matrix $\begin{bmatrix} 1/\sqrt{3} & (1+i)/\sqrt{3} \\ (1-i)/\sqrt{3} & -1/\sqrt{3} \end{bmatrix}$ is unitary.

3. The determinant of an orthogonal matrix has one or other of the values ± 1.

4. The modulus of the determinant of a unitary matrix is unity.

3. **Special Subgroups of the Group of Equivalent Transformations.**

The more important types of equivalent transformation PAQ are the following:

(i) *The Collineatory Subgroup*

If $PQ = I$, we may write $P = H$, $Q = H^{-1}$, $|H| \neq 0$. The transformation so derived, HAH^{-1}, is of outstanding importance in the pure theory of matrices. For if t is a positive integer we have

$$\begin{aligned}(HAH^{-1})^t &= HAH^{-1}HAH^{-1}\ldots HAH^{-1} \\ &= HAIAI\ldots AH^{-1} = HA^tH^{-1}. \quad . \quad (4)\end{aligned}$$

Also, by the reversal law, $(HAH^{-1})^{-1} = HA^{-1}H^{-1}$, provided that A is non-singular. It follows without difficulty from these two properties that if $f(A)$ is a rational function of A with scalar coefficients then

$$f(HAH^{-1}) = Hf(A)H^{-1}. \quad . \quad . \quad . \quad . \quad . \quad (5)$$

The importance of investigating a canonical form for this transformation is now evident; for once we are in possession of such a form $B = HAH^{-1}$, preferably diagonal, or nearly so, then the properties of matrix functions $f(A)$ can be studied by way of the simpler forms $f(B)$.

The transformation HAH^{-1}, which for geometrical reasons to be set forth later will be called the transformation of *similarity*, or the *collineatory* transformation, possesses group properties. Let

$$B = HAH^{-1}, \quad C = KBK^{-1}, \quad |H| \neq 0, \quad |K| \neq 0.$$

Then

$$C = KHAH^{-1}K^{-1} = SAS^{-1}, \text{ where } S = KH, \ |S| = |K||H| \neq 0.$$

Thus similarity is a transitive property. When $B = HAH^{-1}$, B is termed the *transform* of A by H.

(ii) *The Congruent Subgroup*

If $P = Q'$, we have the *congruent* or *correlatory* subgroup, $H'AH$, $|H| \neq 0$. Here again the transitive property holds, for if $B = H'AH$, $C = K'BK$, then $C = K'H'AHK = S'AS$, where $S = HK$, $|S| \neq 0$.

(iii) *The Conjunctive Subgroup*

The matrices here considered have complex elements, and $P = \overline{Q}'$, $|Q| \neq 0$. The subgroup so derived, $B = \overline{H}'AH$, $|H| \neq 0$, is called the *conjunctive* subgroup of the complex group. If $C = \overline{K}'BK$, we have $C = \overline{K}'\overline{H}'AHK = \overline{S}'AS$, where $S = HK$, $|S| \neq 0$; so that the transitive property again holds.

Evidently the conjunctive transformation is an extension of the congruent transformation, into which it merges in the case of real matrices.

(iv) *The Orthogonal Subgroup*

A transformation which is at the same time collineatory and congruent is said to be *orthogonal*. In this case $PQ = I$, $P = Q'$. We have therefore the transformation $B = H'AH$, where $H' = H^{-1}$, $|H| \neq 0$. The transitive property is readily verified.

(v) *The Unitary Subgroup*

If, on taking the complex analogue of the orthogonal case above, we combine the collineatory and the conjunctive conditions, we have the *unitary* subgroup. In this case $B = \overline{H}'AH$, $H' = \overline{H}^{-1}$, $|H| \neq 0$. The transitive property is again easily verified.

EXAMPLES

1. The congruent transforms of a symmetric matrix are symmetric. Hence the orthogonal transforms of symmetric matrices are symmetric.

2. The conjunctive transforms of an Hermitian matrix are Hermitian. Hence the unitary transforms of Hermitian matrices are Hermitian.

3. The congruent transforms of a skew symmetric matrix are skew symmetric.

4. The conjunctive transforms of a skew Hermitian matrix are skew Hermitian.

4. Quadratic and Bilinear Forms Associated with the Subgroups.

Each of the above subgroups owes much of its importance to the fact that it is closely connected with a certain bilinear or quadratic form. In § 1 of Chapter III we saw that the main equivalent group came into view through a consideration of linear transformations $u = vP$, $x = Q\xi$ which were independent of each other. The sets of variables u_i and x_j were supposed to belong to distinct realms. In practice this is by no means the common case: and one must consider different types of relation which may subsist between the variables. We shall take them in order.

(i) *Contragredient Variables. The Collineation*

Let u, x denote the two sets of n variables in the bilinear form

$$f = uAx = \sum_{i,j=1}^{n} u_i a_{ij} x_j . \qquad \qquad (6)$$

Let a new bilinear form $vHAH^{-1}y$ be derived from f by the linear transformations

$$u = vH, \quad x = H^{-1}y, \quad |H| \neq 0. \qquad \qquad (7)$$

Such variables u, x, since they undergo opposite transformations, are said to be *contragredient*, and the equation $f = uAx = 0$ belongs to a *collineation*. (See p. 40, *infra.*) It is this bilinear form f which is associated with the collineatory subgroup.

From (7) it follows at once that $ux = vy$; in other words, that the inner product $ux = \Sigma u_i x_i$ is an *absolute invariant* of the collineatory group, being exactly equal to the same function of the transformed variables. Briefly, we shall say that ux is *latent* in the group. Conversely, when $ux = vy$ identically, the variables u and x are contragredient; for if in general we have $uAx = vPAQy$, so that $u = vP$, $x = Qy$, and if further $ux = vy$ identically, then $vPQy = vy$ and so $PQ = I$. (Cf. *Invariants*, p. 148.)

(ii) *Cogredient Variables. The Correlation*

Suppose on the other hand that we make the variables x, y in a bilinear form

$$\Gamma = y'Ax = \sum_{i,j} y_i a_{ij} x_j \qquad \qquad (8)$$

undergo the *same* transformation: let us say $x = H\xi$, $y = H\eta$. Then

$$y'Ax = (H\eta)'A(H\xi) = \eta'H'AH\xi = \eta'B\xi, \quad . \quad . \quad (9)$$

where $B = H'AH$; and here we recognize the congruent or correlatory subgroup. Such variables x and y are said to be *cogredient*; they are illustrated by points in analytical geometry. The equation $\Gamma = 0$ is called a *correlation*.

If, in particular, A is symmetric and y is identical with x, then Γ becomes a *quadratic form* $x'Ax$. After linear transformation this in turn becomes $\xi'H'AH\xi$, where the new matrix $H'AH$ is also symmetric (since $(H'AH)' = H'AH$), as is otherwise obvious.

(iii) *Conjunctive Transformations*

If the variables x, $\bar{x}$ of a Hermitian form undergo transformation according to

$$x = H\xi, \quad \bar{x} = \bar{H}\bar{\xi}, \quad |H| \neq 0, \quad . \quad . \quad . \quad (10)$$

the resulting form is

$$\bar{x}'Ax = \bar{\xi}'\bar{H}'AH\xi = \bar{\xi}'B\xi, \quad . \quad . \quad . \quad . \quad (11)$$

where $B = \bar{H}'AH$. Thus A has been transformed into B by a conjunctive transformation; and we note that the conjunctive transformation is associated with Hermitian forms exactly as the congruent transformation is with quadratic forms, the latter case, for real transformations, being indeed a special instance of the former.

(iv) *Unitary and Orthogonal Transformations*

If we premultiply a vector $x = \{x_1, x_2, \ldots, x_n\}$, having complex elements in a given field $\mathcal{F}$, by its transposed complex conjugate $\bar{x}'$, we derive the matrix product

$$\bar{x}'x = \bar{x}_1x_1 + \bar{x}_2x_2 + \ldots + \bar{x}_nx_n, \quad . \quad . \quad . \quad (12)$$

a scalar number which is essentially *real* and *non-negative*, since it is evidently a sum of squares of real numbers, the moduli of the x_i in fact, and which is indeed only zero when x is null. This fundamental Hermitian inner product of $\bar{x}$ and x is often called the *norm* of the complex vector x. (It is a Hermitian form with unit matrix.) The square root of the norm, taken with positive sign, $(\bar{x}'x)^{\frac{1}{2}}$, is sometimes denoted by $|x|$; it is an obvious generalization of the modulus of a complex number.

A vector a such that $|a| = 1$ is said to be *normalized*: for example, $x/|x|$ and $\bar{x}/|x|$ are normalized vectors derived from x and $\bar{x}$ respectively. All vectors except the null vector can be thus normalized.

EXAMPLES

1. Find the norm of the vector pencil $[\lambda|x_i| + |y_i|]$, where λ is scalar and real.

$$[\lambda^2 \sum_i |x_i|^2 + 2\lambda \sum_i |x_i y_i| + \sum_i |y_i|^2.]$$

2. Establish the Schwarzian inequality $(\Sigma |x_i y_i|)^2 \leqslant (\Sigma |x_i|^2)(\Sigma |y_i|^2)$.

[This is the condition that the above quadratic in λ should not be negative. Equality occurs only in the case when the vector x is a scalar multiple of y.]

3. Prove that if q be a normalized vector, and x is an arbitrary vector, then $|q| \equiv (\bar{q}'q) = 1$, $|(\bar{q}'x)| \leqslant |x|$.

Fundamental Latent Forms.

It will now be shown that the inner product $\bar{x}'x$ is latent in the unitary group. For if x is transformed to ξ by a non-singular transformation H we must have

$$x = H\xi, \quad \bar{x} = \bar{H}\bar{\xi}, \quad \bar{x}' = \bar{\xi}'\bar{H}'. \qquad \ldots \quad (13)$$

Hence

$$\bar{x}'x = \bar{\xi}'\bar{H}'H\xi. \qquad \ldots\ldots \quad (14)$$

But if H is unitary $\bar{H}'H = I$, and so $\bar{x}'x = \bar{\xi}'\xi$; that is, $\bar{x}'x$ is latent in the group. Conversely we may show that if $\bar{\xi}'\xi$ is equal to $\bar{x}'x$ identically for all values of x, then H must be a unitary matrix; for

$$\bar{x}'x = \bar{\xi}'\bar{H}'H\xi = \bar{\xi}'\xi, \qquad \ldots\ldots \quad (15)$$

and consequently $\bar{H}'H = I$ since the ξ_i are arbitrary.

If the elements and variables concerned in the above are real, the case reduces to the orthogonal case. This is evident on inspection. The form that is latent in the orthogonal group is then seen to be the quadratic form $x'x$ or Σx_i^2. There is also a complex orthogonal case in which $x'x$ is invariant where x_i is complex, but, though many of its properties are easily shown to be the same as those of the real orthogonal case, it is relatively unimportant.

The latency of Σx_i^2 gives one of several reasons for the choice of the word "orthogonal": for the sum of squares in question represents, in a suitable system of rectangular (that is, *orthogonal*) Cartesian axes, the square of the distance of the point x from the origin, and remains invariant for a change of axes, provided that they remain rectangular, with fixed origin.

5. Geometrical Interpretation of the Collineation.

Let x denote a point in space of $n-1$ dimensions, the n elements x_i of the matrix x being homogeneous co-ordinates of the point referred to a given simplex or frame of reference. (*Invariants*, pp. 86, 293.) Since the ratios of the co-ordinates are sufficient to determine the point, it follows that the same point is also given by

$$\lambda x = \lambda\{x_1, x_2, \ldots, x_n\}, \quad \lambda \neq 0. \quad . \quad . \quad . \quad . \quad (16)$$

We shall exclude for the moment the case when every component is zero. If now A is a non-singular square matrix of order n, there is a certain point $y = Ax$ which is unique; while reciprocally we have $x = A^{-1}y$, so that x is also uniquely determined if y is given. The one-one correspondence thus set up, between what we may call object points x and their images y with respect to A, constitutes a collineation.

Consider now the case when $x = 0$. Obviously $Ax = 0$ also. Conversely if $y = Ax = 0$, and A is non-singular, then the only admissible solution of the n equations involved is $x = 0$. (Cf. p. 10.) Hence in the non-singular case if x is zero so is y, and conversely.

The case is different for singular collineations, that is, those for which $y = Ax$, but $|A| = 0$. There is then no reciprocal or inverse collineation, and so there are actually object points which possess no images; these are in fact determined by values of x_i, not all zero, which satisfy $\Sigma a_{ij}x_j = 0$. There exists therefore a certain locus of points x which may be said to be *annihilated* by A. If x and z are two such points, we have $Ax = 0$, $Az = 0$, so that $A(x + \lambda z) = 0$, where λ is an arbitrary scalar. But $x + \lambda z$ denotes a point in the *same straight line* as x and z. Hence if two points are annihilated, so are all points in their common straight line. Similarly if three coplanar but non-collinear points are annihilated, so is their common plane: and so on. If the linear locus which is annihilated is of $\alpha - 1$ dimensions, the matrix A may be said (following Sylvester) to be of *nullity* α; and, in fact, for a matrix of order n the nullity α, it will be found, is connected with the rank r by the relation $r = n - \alpha$.

6. The Poles and Latent Points of a Collineation.

When object and image point coincide, the point is said to be a *latent point* of the collineation and of the matrix A. If x is a latent point then, in order that x and y may coincide, we must have $y = \lambda x$, that is, $y_i = \Sigma a_{ij}x_j = \lambda x_i$. The condition for latency reduces therefore to that of the consistency of n homogeneous linear equations in the x_i;

and when the x_i are eliminated we obtain the *characteristic equation* of A, namely,

$$|A-\lambda I| = \begin{vmatrix} a_{11}-\lambda & a_{12} & a_{13} & \dots & a_{1n} \\ a_{21} & a_{22}-\lambda & a_{23} & \dots & a_{2n} \\ \cdot & \cdot & \cdot & \cdot & \cdot \\ a_{n1} & a_{n2} & a_{n3} & \dots & a_{nn}-\lambda \end{vmatrix} = 0. \quad (17)$$

The determinant $|A-\lambda I|$ is called the *characteristic determinant* of A: when it is expanded in powers of λ we obtain a polynomial, of order n in λ, namely, the *characteristic function*

$$\phi(\lambda) = (-)^n[\lambda^n - p_1\lambda^{n-1} + p_2\lambda^{n-2} - \dots + (-)^n p_n], \quad (18)$$

where p_r is the sum of the diagonal minors of order r of A, and p_n is the determinant $|A|$ itself. (*Invariants*, p. 98.) The equation has n roots $\lambda_1, \lambda_2, \dots, \lambda_n$, real, complex, or zero, called the *latent roots* of the matrix A. For each distinct value of λ_i the condition (17) is satisfied; and a set of x_i can be found which are not all zero. If $\lambda_i \neq 0$ this gives a vector y proportional to x, and therefore a *latent point* x. If $\lambda_i = 0$ it gives a zero y, so that x is an object point annihilated by A. We shall call the point x a *pole* of the collineation whenever $y = \lambda_i x$, whether $\lambda_i = 0$ or $\lambda_i \neq 0$: so that the poles include all latent points, together with all points annihilated by A; and no others. Each zero root λ_i that may exist is, however, called a latent root of the matrix A.

EXAMPLES

1. The poles of the collineation $y = Ax$, where $A = \begin{bmatrix} \lambda & \cdot & \cdot \\ \cdot & \mu & \cdot \\ \cdot & \cdot & \nu \end{bmatrix}$ and $\lambda\mu\nu \neq 0$, are $\{1, 0, 0\}$, $\{0, 1, 0\}$, $\{0, 0, 1\}$.

2. The matrix $\begin{bmatrix} \cdot & 1 & 2 \\ \cdot & \cdot & 3 \\ \cdot & \cdot & \cdot \end{bmatrix}$ has no latent points, but has one pole $\{1, 0, 0\}$.

3. *Every matrix of order* n *has at least one pole.*

[Cf. p. 29.

7. **Change of Frame of Reference.**

The collineatory transformation of the matrix A itself may also be interpreted from the present standpoint. We have been considering A as a kind of optical instrument which transforms object points x into image points y, where both x and y are referred to the *same* frame of reference. Suppose now that a new frame is chosen, defined by the linear transformation

$$\xi = Hx, \quad \xi_i = \Sigma h_{ij}x_j, \qquad i = 1, 2, \dots, n, \quad . \quad (19)$$

giving the new co-ordinates ξ in terms of the x. For the new to be a proper, and not a degenerate, frame the matrix H must be non-singular, so that $|H| \neq 0$. If now η denotes any point y referred to the new frame, then $\eta = Hy$, or, since H is non-singular, $y = H^{-1}\eta$. But if x and y are object and image we have $y = Ax$. Hence

$$\eta = Hy = HA\,x = HAH^{-1}\xi = B\,\xi, \quad . \quad . \quad . \quad (20)$$

where $B = HAH^{-1}$. Thus the original geometrical collineation of matrix A, when referred to a new frame of reference through a transformation of the variables with matrix H, becomes a collineation of matrix HAH^{-1}.

The properties set for demonstration below, though not difficult to verify, are important.

EXAMPLES

1. If B is the transform of A by H, A is the transform of B by H^{-1}.

2. *The matrices* A *and* B = HAH^{-1} *have the same characteristic equation and therefore the same latent roots.*

[Since $B = HAH^{-1}$, then $B - \lambda I = H(A - \lambda I)H^{-1}$. Forming the determinants of both sides we have $|B - \lambda I| = |H|\,|A - \lambda I|\,|H|^{-1} = |A - \lambda I|$; and the results follow.]

3. *If* x *is a latent point, or a pole, of* A *and if* H *transforms* x *into* y, *then* y *is a latent point, or a pole, of* B.

[Since $Ax = \lambda x$ and $y = Hx$, therefore $AH^{-1}y = \lambda H^{-1}y$ and so $HAH^{-1}y = \lambda y$. Thus $By = \lambda y$, which proves the theorem. It is instructive, as showing the power of matrix algebra, to rewrite this proof in its full prolixity for the case when $n = 3$.]

8. Alternative Geometrical Interpretation.

In the interpretation just given we transformed the frame of reference, not the points under view. We might equally well have kept the original frame of reference while transforming the points. Then the three equations $y = Ax$, $\xi = Hx$, $\eta = Hy$ would determine, from a given point x, two new points ξ and y: and again from y a further point η. These four points are in general distinct, and form a quadrilateral, which is usually skew if $n > 3$. The single arrow in the figure denotes the operation A changing x to y: the double arrow denotes H. Still another operation is suggested by the remaining segment from ξ to η; and

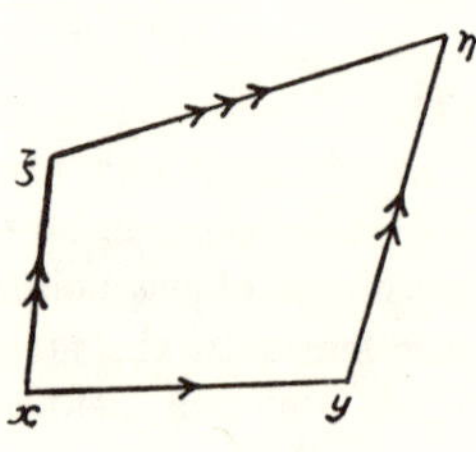

the algebra at once shows that this is the operation $B = HAH^{-1}$, for we have $\eta = HAH^{-1}\xi$.

The diagram shows vividly why B is called the *transform* of A by H; for the operation A is palpably transformed into B by H. Again, supposing y coincides with x, then η (uniquely given by Hx) must coincide with ξ. Hence the matrix H transforms latent points of A into latent points of B, as has already been proved in Example 3 above. Again, if $x \neq 0$, $y = 0$, then $\xi \neq 0$, $\eta = 0$. In this case neither point y nor η will exist in the figure: and x, ξ are both poles of their matrices.

Once more, let a sequence of points be taken such that each is the image of its immediate predecessor with respect to A; such a sequence may be denoted by $x, Ax, A^2x, \ldots$. If these are treated as object points for the matrix H, their images will be $\xi, B\xi, B^2\xi, \ldots$; for $HA^rx = HA^rH^{-1}\xi = B^r\xi$, by (4) of § 3, p. 35. Further, if the matrices are of order n, the figure is situated in space of $n - 1$ dimensions, in which case, to anticipate for a moment the topic of the next chapter, n points at most are *linearly independent*. Consequently the first $n + 1$ terms of the sequence $x, Ax, A^2x, \ldots$, and possibly still fewer, are connected by a relation

$$(A^n + q_1A^{n-1} + q_2A^{n-2} + \ldots + q_nI)\,x \equiv \psi(A)x = 0, \quad (21)$$

where the q_i are scalar constants. Thus each point x has a *characteristic function* with respect to a matrix A, that is, a polynomial in A of lowest degree with scalar coefficients, $\psi(A)$, such that $\psi(A)\,x = 0$. Such functions, which also exist, as postmultipliers, for *vectors of the first kind*, will be considered in detail in the next chapter.

9. **The Cayley-Hamilton Theorem.**

Cayley, in his original memoir of 1858, enunciated but did not completely prove a fundamental theorem in matrices, namely, that any square matrix A of order n, whether singular or not, satisfied the identity $\phi(A) = 0$, where $\phi(\lambda) = |\,A - \lambda I\,|$; in other words, *if in the expansion of the characteristic determinant of* A *we replace* λ *by* A, *the resulting matrix polynomial is null.* Various proofs of this remarkable theorem exist; some of the simplest depend on the adjoint of $A - \lambda I$, which will of course be a λ-matrix. (Cf. *Invariants*, p. 100.) For we have, as in Chapter I,

$$(A - \lambda I)\,\mathrm{adj}\,(A - \lambda I) = \phi(\lambda)I. \quad . \quad . \quad . \quad (22)$$

But, as has already been observed in Chapter III, p. 23, scalar polynomials in A behave like ordinary polynomials, and so, by the remainder theorem,

$$\phi(A) - \phi(\lambda)I \equiv 0, \quad \text{mod } A - \lambda I. \quad . \quad . \quad . \quad (23)$$

Comparison of (22) and (23) shows that $\phi(A)$ itself, considered as a λ-matrix, must be divisible by $\lambda I - A$; but λ is arbitrary, and $\phi(A)$ is independent of λ. Hence $\phi(A)$ must be zero, which proves the theorem.

The circumstances under which A may satisfy a characteristic equation of *lower degree than n* will receive consideration in the two following chapters.

10. **Historical Note.**—The Hermitian bilinear form was first treated by Hermite in 1854; *J. für Math.*, **47**, pp. 343–68, or *Œuvres*, I, 234–63. Traces of it appear in the letters of Hermite to Jacobi of slightly earlier date.

Cayley's remark concerning the Cayley-Hamilton theorem is curious. He says (*Collected Works*, II, p. 483): " I have not thought it necessary to undertake the labour of a formal proof of the theorem in the general case of a matrix of any degree." He proposed to write the theorem in a notation equivalent to $| a_{ij}I - \delta_{ij}A | = 0$. Hamilton's title to the theorem is based on the fact that he had previously established the existence of characteristic equations for quaternions (*Lectures on Quaternions* (Dublin, 1853), 566–7). The terms *latent root* and *latent point* are due to Sylvester: see for example a characteristic metaphor on latency (*Collected Works*, IV, 110).

CHAPTER V

A Rational Canonical Form for the Collineatory Group

In the present chapter we shall consider a square matrix A with constant elements belonging to a prescribed field, the object in view being to establish a *rational* canonical form through collineatory transformations, that is, to reduce A to a certain simple form $B = HAH^{-1}$ by processes which are entirely rational in the field. The fundamental idea throughout will be that of the *linear independence* of vectors in the field.

1. Linear Independence of Vectors in a Field.

A set of vectors $u, v, \ldots, w$ in the field $\mathcal{F}$, where $u = [u_1, u_2, \ldots, u_n]$, will be said to be *linearly dependent in* $\mathcal{F}$ if there exists a relation

$$\Sigma \alpha u = \alpha u + \beta v + \ldots + \gamma w = 0, \quad . \quad . \quad . \quad (1)$$

where the coefficients $\alpha, \beta, \ldots, \gamma$ are numbers not all zero, belonging to $\mathcal{F}$. If no such relation exists, the vectors will be termed *linearly independent* in $\mathcal{F}$. The n components of the vector u may also be regarded as coefficients of a linear form

$$\xi = u_1 x_1 + u_2 x_2 + \ldots + u_n x_n = ux \quad . \quad . \quad . \quad (2)$$

in an arbitrary set of variables $x = \{x_1, x_2, \ldots, x_n\}$, and similarly for $\eta = vx, \ldots, \omega = wx$: we may then replace (1) by the alternative relation

$$\alpha ux + \beta vx + \ldots + \gamma wx = 0, \quad \{x\}: \quad . \quad . \quad . \quad (3)$$

for evidently (1) stands for n ordinary scalar relations $\Sigma \alpha u_i = 0$, $(i = 1, 2, \ldots, n)$, involving the various components u_i; and (3) means the same, the notation $\{x\}$ implying that it is an identity for all values of the x_i. Also, in respect of dependence or independence in the field, the forms $\xi, \eta, \ldots, \omega$ behave exactly as the vectors $u, v, \ldots, w$. Although the elements $u_i, v_i, \ldots, \alpha, \beta, \ldots$ are members of $\mathcal{F}$, it is at present immaterial whether the variables x_i belong to $\mathcal{F}$ or not.

EXAMPLES

1. The null vector $u = [0, 0, \ldots, 0]$ is linearly dependent.

2. The n vectors $i_1 = [1, 0, 0, \ldots, 0]$, $i_2 = [0, 1, 0, \ldots, 0], \ldots,$ $i_n = [0, 0, \ldots, 0, 1]$, that is, the n vectors which form the rows of the unit matrix I, are linearly independent.

These vectors form a *basis*, in that any other vector of order n in the field can be expressed linearly in terms of them as $\sum_{k=1}^{n} u_k i_k$, the components u_k being numbers of the field.

3. The two vectors u and $v = [\alpha u_1, \alpha u_2, \ldots, \alpha u_n]$ are linearly dependent.

4. Any $n + 1$ vectors of order n in a field must be linearly dependent.

5. Any vector u of order n can be expressed linearly in terms of n given linearly independent vectors $v^{(1)}, v^{(2)}, \ldots, v^{(n)}$.

[The n equations $u_j = \Sigma \alpha_i v_{ij}$, $(j = 1 \text{ to } n)$, can be solved for the α_i, if $|v_{ij}| \neq 0$. Thus $\alpha_j = \Sigma u_i v^{ji}$. Take the rows of $[v_{ij}]$ to be the given vectors $v^{(i)}$ and the result $u \equiv \alpha_1 v^{(1)} + \ldots + \alpha_n v^{(n)}$ follows.]

6. If r vectors of the nth order, $\xi_1, \xi_2, \ldots, \xi_r$ are given as linearly independent in $\mathcal{F}$, then $n - r$ further vectors exist such that all n vectors are linearly independent.

[If $|X_r|$ is a non-zero minor determinant of order r belonging to the $r \times n$ matrix $\{\xi_1, \xi_2, \ldots, \xi_r\}$ having for ith row the vector ξ_i, then the square matrix $Y = \begin{bmatrix} X_r & \cdot \\ \cdot & I_{n-r} \end{bmatrix}$ is non-singular, I_{n-r} being a complementary unit matrix of order $n - r$. The last $n - r$ rows of Y furnish the desired vectors. This is the simplest but by no means the unique solution.]

7. If $c_1, c_2, \ldots, c_n$ are n different numbers belonging to $\mathcal{F}$, the n vectors $u_i = [1, c_i, c_i^2, c_i^3, \ldots, c_i^{n-1}]$ are linearly independent.

8. If a determinant Δ vanishes, its rows (or columns) are linearly dependent.

[If the vectors $u, v, \ldots, w$ are the rows, these are linearly dependent if there exists a relation

$$\lambda u_i + \mu v_i + \ldots + \nu w_i = 0, \qquad i = 1, 2, \ldots, n,$$

where $\lambda, \mu, \ldots, \nu$ are not all zero. But the vanishing of Δ is the condition (p. 29) that these homogeneous equations should possess a solution.]

2. The Reduced Characteristic Function of a Vector.

The matrix product uA, formed from a vector u and a square matrix A of the same order, yields a new vector $\hat{u} = [\hat{u}_1, \hat{u}_2, \ldots, \hat{u}_n]$. More generally if $\psi(A)$ is any polynomial function of A, then $u\psi(A)$ is again a vector. Suppose then that we have a certain non-zero vector u in the field $\mathcal{F}$; we may proceed to construct, with respect to a given matrix A in the field, a sequence of vectors

$$u,\ uA,\ uA^2, \ldots, uA^{m-1}, \qquad \ldots \ldots \quad (4)$$

the sequence being continued just so long as the vectors comprised in it remain linearly independent. Since n is the greatest number of vectors which can be thus attained we must have $0 < m \leqslant n$. Again, since the sequence terminates because the next vector uA^m is linearly related to the preceding m vectors, there must be a relation

$$c_0uA^m - c_1uA^{m-1} - c_2uA^{m-2} - \ldots - c_mu = 0, \quad . \quad (5)$$

where the c_i belong to $\mathcal{F}$, and are not all zero. Also c_0 is not zero, for if it were the vectors in (4) would not be independent. Hence without loss of generality c_0 may be taken to be unity; then, in virtue of the distributive law, the factor u may be taken to the left outside a bracket, and an annihilating matrix polynomial U appears in the form

$$uU \equiv u(A^m - c_1A^{m-1} - c_2A^{m-2} - \ \ldots - c_mI) = 0. \quad (6)$$

The matrix polynomial U is called the *Reduced Characteristic Function* (briefly the R.C.F.) of A relative to the vector u in the field $\mathcal{F}$. Where no misunderstanding is likely to arise we shall call it the R.C.F. of u. The properties of A are closely bound up with the nature of the various polynomials U.

3. Fundamental Theorem of the Reduced Characteristic Function.

Theorem I.—*For a given arbitrary non-zero vector* u *in the field* $\mathcal{F}$ *of a square matrix* A *of the same order, the R.C.F.* U *is unique, and is a divisor of any other polynomial* ψ(A) *in the field such that* uψ(A) = 0.

Proof.—If U is not a divisor of ψ, then a non-zero polynomial $g(A)$ exists, as in § 5, p. 23, having coefficients in $\mathcal{F}$, and of order less than m, such that $U(A)k(A) - \psi(A)h(A) = g(A)$. Pre-multiplying by u we have $uU \cdot k(A) - u\psi \cdot h(A) = ug(A)$. Since both uU and $u\psi$ are null vectors, so is $ug(A)$. But this contradicts the fact that U is the polynomial of lowest degree in A which annihilates u. Hence U is a factor of ψ.

Again, if u were to possess two different R.C.F.'s U and U_1, of the same degree m, then the non-zero polynomial $U - U_1$, of degree less than m (since highest terms cancel), would satisfy $u(U - U_1) = 0$, which is once more contrary to hypothesis. Hence the R.C.F. of u is unique.

Definition of Grade of a Vector.—*The degree* m *of the R.C.F.* U *is called the grade of the vector* u *with respect to the matrix* A *in the field* $\mathcal{F}$.

All possible vectors u, v, ... of the nth order in $\mathcal{F}$ have their various grades and R.C.F.'s with respect to A. The significance of these differing grades in relation to the invariant properties and the canonical form of A will duly appear.

Consider next the n vectors

$$\begin{aligned} i_1 &= [1,\ 0,\ 0, \ldots,\ 0], \\ i_2 &= [0,\ 1,\ 0, \ldots,\ 0], \\ &\cdots \\ i_n &= [0,\ 0,\ \ldots,\ 0,\ 1], \end{aligned}$$

in terms of which as basis, as we have mentioned in Example 2 of § 1, any vector u can be resolved into components

$$u = u_1 i_1 + u_2 i_2 + \ldots + u_n i_n.$$

Let $U_1, U_2, \ldots, U_n$ denote the n R.C.F.'s of these n vectors i_ν, and let $\psi(A)$ be the L.C.M. of the polynomials U_ν. Exactly as in ordinary scalar algebra $\psi(A)$ will be unique, will have unity for coefficient of its highest power of A, and will contain each U as a factor. It follows that

$$u\psi = u_1 i_1 \psi + u_2 i_2 \psi + \ldots + u_n i_n \psi = 0; \quad . \quad . \quad (7)$$

and this is true for every vector in the field. Now by matrix multiplication we have

$$\psi(A) = I\psi = \begin{bmatrix} i_1 \\ i_2 \\ \vdots \\ i_n \end{bmatrix} \psi = \begin{bmatrix} i_1\psi \\ i_2\psi \\ \vdots \\ i_n\psi \end{bmatrix} = \begin{bmatrix} 0 \\ 0 \\ \vdots \\ 0 \end{bmatrix} = 0. \quad . \quad (8)$$

Hence $\psi(A)$ vanishes identically, and we shall show in fact that

$$\psi(A) = A^p - a_1 A^{p-1} - a_2 A^{p-2} - \ldots - a_p I = 0, \quad 0 < p \leqslant n, \quad (9)$$

which is called the *Reduced Characteristic Equation* of the matrix A in the field $\mathcal{F}$. By Theorem I, since $u\psi = 0$, and since for any vector u in the field $uU = 0$, where U is the R.C.F. of u, it follows that ψ contains U as a factor, and is thus the L.C.M. not merely of the U_ν but also of all the R.C.F.'s of vectors of order n in the field. Conversely, too, A cannot satisfy any equation $\chi(A) = 0$ of degree lower than p: for if it did, then $u\chi$ would be zero for all vectors u, so that χ would be, and ψ could not be, the *least* common multiple of all the U_ν, contrary to hypothesis. Hence the assertion is proved.

The Reduced Characteristic Function ψ(A) *of a matrix* A *is a factor of the Characteristic Function* ϕ(A), as defined in Chapter IV. (This includes the possibility that ψ and ϕ are identical—the *elementary* case.) For if this were not so we should find, precisely as in the first part of the present section, a polynomial $g(A)$ of lower degree than either, such that $g(A) = 0$, contrary to what has just been proved.

4. A Rational Canonical Form for Collineatory Transformations.

Theorem II.—*By rational collineatory transformations within a given field* $\mathcal{F}$ *containing all the elements* a_{ij} *of a square matrix* A *of order* n, *a canonical form*

$$B = \begin{bmatrix} B_p & . & . & \dots & . \\ . & B_q & . & \dots & . \\ . & . & B_s & \dots & . \\ . & . & . & . & . \\ . & . & . & \dots & B_t \end{bmatrix} = HAH^{-1} \quad . \quad . \quad (10)$$

may be found, such that each diagonal submatrix B_i *is of the type, e.g.*,

$$B_p = \begin{bmatrix} . & 1 & . & \dots & . \\ . & . & 1 & \dots & . \\ . & . & . & . & . \\ . & . & . & \dots & 1 \\ a_p & a_{p-1} & a_{p-2} & \dots & a_1 \end{bmatrix}, \quad n \geqslant p \geqslant q \geqslant \dots \geqslant t, \quad (11)$$

all the elements of B *and* H *belonging also to the field.*

Proof.—Of the vectors of order n in $\mathcal{F}$, there will be vectors of highest grade, p, with respect to A. Let one of these, u say, be taken, and let the chain of (linearly independent) vectors u, uA, $uA^2, \dots,$ uA^{p-1} be formed. Let these p vectors in order be taken to be the first p rows of the premultiplying n-rowed matrix H, and let sets of n variables $x = \{x_1, x_2, \dots, x_n\}$, y, ξ, η, be introduced by the collineatory relations of the last chapter,

$$y = Ax, \quad \xi = Hx, \quad \eta = Hy, \quad \text{so that } \eta = HAH^{-1}\xi. \quad (12)$$

We have then, corresponding to the first p rows of Hx,

$$\left.\begin{aligned} \sum_{k=1}^{n} u_k x_k = ux \quad &= \xi_1, \\ \eta_1 = uAx \quad &= \xi_2, \\ \eta_2 = uA^2x \quad &= \xi_3, \\ . \quad . \quad . \quad . \quad &. \quad . \quad . \\ \eta_{p-1} = uA^{p-1}x &= \xi_p, \\ \eta_p = uA^px \quad &= a_p\xi_1 + a_{p-1}\xi_2 + \dots + a_1\xi_p, \\ &0 < p \leqslant n. \end{aligned}\right\} \quad (13)$$

The last of the above relations is the result of postmultiplying the

R.C.F. of u by x, transferring all but the first term to the right side of the identity and substituting the appropriate ξ_i's. Thus we have the first p of the η_i linearly in terms of the ξ_i, and this (as in § 5, p. 14) yields the desired form B_p for the first p rows of the canonical matrix B. If $p = 1$, we have $\eta_1 = a_1\xi_1$, so that $B_1 = a_1$; or again, in the elementary case $p = n$, the reduction of A is complete with $B_p = B$.

If, however, $p < n$, there will be other vectors v linearly independent of the preceding vectors $u, uA, \ldots, uA^{p-1}$, and we may form a chain $v, vA, \ldots, vA^{q-1}$, continuing just so long as $u, uA, \ldots, uA^{p-1}$, $v, vA, \ldots, vA^{q-1}$ are linearly independent. For the different vectors v remaining there may possibly be more than one value of q; if that be so, we choose a v such that q takes the highest value. Then if vA^q be the first vector which is linearly dependent on the $p + q$ preceding vectors, we have

$$vA^q = b_1vA^{q-1} + b_2vA^{q-2} + \ldots + b_qv + k_1uA^{p-1} + \ldots + k_pu, \quad (14)$$

which we may write $vV = uK$, where

$$\left.\begin{aligned} V &= A^q - b_1A^{q-1} - b_2A^{q-2} - \ldots - b_qI, \\ K &= k_1A^{p-1} + k_2A^{p-2} + \ldots + k_pI, \end{aligned}\right\} \quad . \quad . \quad (15)$$

where $p \geqslant q > 0$, and all of b_i and k_i belong to $\mathcal{F}$. By the mode of construction of the chain no relation $vV_1 = uK_1$ exists, where V_1 is a polynomial of degree *less* than q. Had we taken instead of v any vector of the form $\bar{v} = v - uQ$, where Q is an arbitrary matrix polynomial in A, a result similar to (15) would have been obtained,

$$\bar{v}V = uM, \quad \text{where } M = K - QV; \quad . \quad . \quad . \quad (16)$$

and the new vector $\bar{v}$ could satisfy no relation $\bar{v}V_1 = uM_1$ involving a V_1 of lower degree than V, for then on resubstitution v would do so too: which is not the case. Hence (16) gives $\bar{v}A^q$ in terms of the $p + q$ linearly independent vectors

$$u, uA, uA^2, \ldots, uA^{p-1}, \bar{v}, \bar{v}A, \bar{v}A^2, \ldots, \bar{v}A^{q-1}. \quad . \quad (17)$$

Now Q is arbitrary: in keeping with (16) let it be the quotient when K is divided by V, so that M is the remainder and is therefore a polynomial of degree less than q. We shall prove that M is identically zero.

In fact if M is identically zero, then $\bar{v}V = 0$, and V is the R.C.F. of $\bar{v}$. If, however, $M \neq 0$, let P be the R.C.F. of $\bar{v}$, so that $\bar{v}P = 0$, and further let Q_1 be the quotient and N the remainder when P is divided by V, so that $P = VQ_1 + N$, with N of lower degree than V. But then

$$0 = \hat{v}P = \hat{v}(VQ_1 + N) = uMQ_1 + \hat{v}N,$$

since $\hat{v}V = uM$. The degree of P is equal to or less than p, that of V is q, and of M is less than q. Hence that of MQ_1 is less than $q + p - q$, and therefore less than p. Thus N cannot vanish, otherwise u would be annihilated by MQ_1 which is of degree less than p.

We have thus been led to a relation similar to (15), but with N of lower degree than V, which is contrary to hypothesis. Hence M can only be zero, and V is the R.C.F. of $\hat{v}$. By taking q further variables,

$$\xi_{p+1} = \hat{v}x, \quad \xi_{p+2} = \hat{v}Ax, \ldots, \quad \xi_{p+q} = \hat{v}A^{q-1}x, \qquad (18)$$

we have a second chain similar to that of (13),

$$\left.\begin{aligned} \hat{v}x &= \xi_{p+1}, \\ \eta_{p+1} = \hat{v}Ax &= \xi_{p+2}, \\ \cdot\ \cdot\ \cdot\ \cdot\ &\cdot\ \cdot\ \cdot\ \cdot \\ \eta_{p+q} = \hat{v}A^q x &= b_q\xi_{p+1} + b_{q-1}\xi_{p+2} + \ldots + b_1\xi_{p+q}, \\ &(0 < q \leqslant p \leqslant n); \end{aligned}\right\} \qquad (19)$$

and (16) shows that all of the $p + q$ vectors η are linearly independent. We thus derive q further rows of the canonical form B, in the shape of a second canonical submatrix B_p.

If $p + q < n$, then vectors w, linearly independent of the $p + q$ vectors already considered, are still outstanding, and we can construct a further chain $w, \ldots, wA^{s-1}$ of vectors, in such a way as to have s as large as possible and to have in all $p + q + s$ linearly independent vectors. This time wA^s is the first vector that is linearly related to the preceding vectors and chains. In place of (15) we have a relation

$$wW = uK_2 + \hat{v}K_3, \qquad (20)$$

where W is a polynomial in A of degree s. By a device analogous to the earlier one, we may use in place of w a modified vector

$$\hat{w} = w - uQ_2 - \hat{v}Q_3, \qquad (21)$$

where Q_2, Q_3 are the quotients when K_2, K_3 are divided respectively by W. This leads as before to $\hat{w}W = 0$; a fresh instalment of transformed variables can be set up, yielding a third canonical submatrix B_s.

No essentially new feature arises in the process of deriving any remaining submatrices B_i; the formation of chains ultimately either exhausts the n rows or arrives at null elements, and the canonical form B is therefore established.

The uniqueness of the form B will come up for consideration later.

5. Properties of the R.C.F.'s of the Canonical Vectors.

We shall now prove a theorem which connects the present investigation with the more general results of the equivalent case given in Chapter III.

Theorem III.—*The several R.C.F.'s* U, V, W, . . . *belonging respectively to the vectors* u, v, w, *. . . *which lead the linearly independent chains in the canonical form* B, *are such that each is either equal to its successor, or contains its successor as a factor.*

Proof.—Consider the vector $u+v$: let its R.C.F. be denoted by U_1, so that $(u+v)U_1 = 0$. As before, on dividing U_1 by U let $U_1 = UQ + R$. Similarly let $U_1 = VP + S$, where R and S are polynomial remainders of degrees less than p and q respectively. Then

$$0 = (u+v)U_1 = u(UQ+R) + v(VP+S) = uR + vS.$$

Again, by (15), such a relation $uR + vS = 0$ is impossible. Hence all the coefficients in R and S are zero: in other words U_1 is a multiple of both U and V. Hence the degree of U_1 is at least p, that of U. But the degree of U_1 is equal to the grade of $u+v$, which cannot exceed p, the grade of u, since u was chosen from vectors of highest grade. Hence the degree of U_1 is exactly p; therefore U_1 can only differ from U by a constant factor. But each has unity for coefficient of A^p. Thus $U_1 = U$; and U contains V as a factor.

A similar argument proves that V contains W, and so on.

Theorem IV.—*The R.C.F. of the vector* u *of the highest grade is identical with the Reduced Characteristic Function of the matrix* A, *and therefore annihilates all vectors.*

Proof.—Since U is a polynomial in A, it commutes with A: also $uU = 0$. Hence $uA^tU = uUA^t = 0$. Similarly $vA^tU = vUA^t = 0$, since $vV = 0$ and $U \equiv 0$, mod V. Hence each of the n linearly independent vectors $u, uA, \ldots, v, vA, \ldots, w, \ldots$ with which we have been dealing is annihilated by U. Hence again any linear combination of these vectors is annihilated, namely,

$$(a_1u + a_2uA + \ldots + a_{p+1}v + \ldots)U = 0.$$

Therefore, by Example 5, p. 46, θ, any vector whatever, satisfies the equation $\theta U = 0$. But *some* vectors u satisfy no such equation of lower order. The polynomial U must therefore be the *Reduced*

* Circumflexes are now dropped as unnecessary.

Characteristic Function of A; and by (8) must vanish identically. Thus

$$U = \psi(A) = A^p - a_1 A^{p-1} - a_2 A^{p-2} - \ldots - a_p I = 0. \quad (22)$$

6. Observations upon the Theorems.

Before we give still further consequences of the theorem that has given us the canonical form (10), it may perhaps be helpful to offer a few comments on the proof. First in regard to the initial choice of a vector of highest grade, in practice *any* vector may first be chosen; for if vectors of higher grade exist they will necessarily reveal themselves in the course of the systematic and finite process of chain-formation. Thus, if at the second stage it were found that the grade of v were greater, then u would be rejected in favour of v: and similarly at other steps. It is even possible for the genuine highest grade vectors to lie concealed until Theorem III is reached: this happens when the polynomial $\psi(\lambda)$ has factors which are rational in the field, as we shall illustrate in Example 1 below.

Again, whatever the matrix A (even if A is null), every vector has a *positive* grade. It is interesting to note that the necessary and sufficient condition for every vector to possess unit grade is that A should be scalar, as the reader will verify by making $p = 1$ in the above investigation. In this case $U = A - a_1 I = 0$, and none of the chain $vA, \ldots$ are independent of v. At the other end of the scale a much more subtle question arises, namely, under what conditions will all non-zero vectors of order n have maximum grade, n? We must leave this question over.

7. Geometrical and Dual Aspect of Theorem II.

Instead of forming a chain of *row*-vectors u, uA, uA^2, ..., we might equally well have started, as remarked in Chapter IV, p. 43, with an arbitrary *column*-vector x, and have formed a chain x, Ax, A^2x, ..., of p terms. This would have led to a canonical form *transposing* the above form B. If we treat the vectors u as primes and the contragredient vectors x as points, we can give a simple geometrical interpretation of the initial reduction in Theorem II.

The bilinear equation

$$uAx \equiv \sum_{i,j=1}^{n} u_i a_{ij} x_j = 0 \quad \ldots \ldots \quad (23)$$

determines a collineation in space of $n - 1$ dimensions. If we regard the n components u_i as current homogeneous prime coordinates, the equation of the arbitrary object-point x is $ux = 0$, while that of its image is $uAx = 0$. Correlatively, if we treat the x_i as current co-ordinates, the equation $ux = 0$ is that of the prime u, while $uAx = 0$ is that of the prime uA.

Suppose now that we follow Sylvester in beginning with an arbitrary object-point x and iterating images Ax, A^2x, . . . until the chain of points becomes linearly dependent. For example, suppose that the points x, Ax, A^2x do not lie in a line but form a triangle; and further that A^3x lies in the plane of the triangle. However many dimensions ($n > 2$) are considered the vector A^3x is linearly related to the three earlier x, Ax, A^2x; let us say

$$t = \alpha x + \beta y + \gamma z, \qquad (24)$$

where $y = Ax$, $z = A^2x$, $t = A^3x$, and α, β, γ are scalar. On premultiplying throughout by A, we find that the next point in the chain is coplanar with the second, third, and fourth, and therefore with the first three. *In fact, the chain never leaves the plane.*

The canonical form B is derived, from this point of view, by taking the points x, Ax, A^2x as three vertices of a frame of reference. It is of course necessary to take n linearly independent points for the complete frame, or simplex. If therefore $n > 3$, a fourth vertex must in the present case be sought outside the plane of the chain of three just found: and this choice interprets the next step in the proof of the theorem.

Correlatively, a chain of row-vectors u, uA, uA^2, . . . , such as we actually used, determines a set of linearly independent primes, which are taken as the first p faces of the simplex; as for example, if $n = 4$, the faces of a tetrahedron. Only if the matrix A is elementary is it possible for a single chain of primes to furnish a complete simplex.

8. The Invariant Factors of the Characteristic Matrix of B.

We have already seen (Ex. 2, p. 42) that the λ-matrix $B - \lambda I$, the characteristic matrix of B, is such that $|B - \lambda I| = |A - \lambda I|$. If we write out the determinant $|B - \lambda I|$ in full and expand it by a Laplacian development, we obtain

$$|B - \lambda I| = |B_p(\lambda)|\,|B_q(\lambda)| \ldots |B_t(\lambda)|,$$

where

$$|B_p(\lambda)| \equiv |B_p - \lambda I| = \begin{vmatrix} -\lambda & 1 & \cdot & \cdots & \cdot \\ \cdot & -\lambda & 1 & \cdots & \cdot \\ \cdot & \cdot & \cdot & \cdots & \cdot \\ \cdot & \cdot & \cdot & \cdots & 1 \\ a_p & a_{p-1} & a_{p-2} & \cdots & a_1 - \lambda \end{vmatrix}. \qquad (25)$$

On expanding the determinant $|B_p(\lambda)|$ in terms of its last row we obtain

$$(-)^p |B_p(\lambda)| = \lambda^p - a_1\lambda^{p-1} - a_2\lambda^{p-2} - \ldots - a_p$$
$$= U(\lambda) = \psi(\lambda), \text{ by Theorem IV.} \quad . \quad (26)$$

Similarly $(-)^q |B_q(\lambda)| = V(\lambda)$, &c. Hence we have at once that the characteristic function of B (or of A),

$$\phi(\lambda) = |A - \lambda I| = |B - \lambda I| = (-)^n U(\lambda)V(\lambda)W(\lambda)\ldots, \quad (27)$$

which gives an alternative proof that the Reduced Characteristic Function $\psi(\lambda)$ or $U(\lambda)$ is a divisor of the Characteristic Function $\phi(\lambda)$: and that $\psi \equiv \phi$ in, and only in, the elementary case.

It is not difficult to identify the polynomials $U(\lambda)$, $V(\lambda)$, . . . , with the invariant factors $E_i(\lambda)$ of the λ-matrix $A - \lambda I$, but in an order reversed from that of Chapter III. We shall show for example that $U(\lambda)$ is the result of dividing $|B - \lambda I|$ by the H.C.F. of the first minors of $|B - \lambda I|$.

Consider the minor of $|B - \lambda I|$ obtained by deleting the ith row and jth column. If the element (ij) is not within one of the submatrices $B_k(\lambda)$, but is one of the extraneous zeros, the corresponding minor is zero. This we see at once by observing that if a unit were inserted at (ij) and $|B - \lambda I|$ expanded as before, the result would be unchanged. Again, if (ij) falls within a submatrix $B_k(\lambda)$ the corresponding minor will evidently be, apart from sign, the co-factor of (ij) in $|B_k(\lambda)|$ itself, multiplied by the continued product of the remaining $|B_\nu(\lambda)|$'s. Now this co-factor in $|B_k(\lambda)|$ will be a polynomial in λ, and the one of lowest degree is clearly that which has as its own diagonal *all of the units in the superdiagonal* of $B_k(\lambda)$; it is, in fact, *the co-factor of the element in the left-hand bottom corner*, and is numerically equal to unity. Further, the longest run of superdiagonal units is in $B_p(\lambda)$, and $|B_p(\lambda)|$, as we have seen, contains all the other $|B_k(\lambda)|$ as factors. We infer that the particular minor of value $1 . V(\lambda)W(\lambda)\ldots$ is the H.C.F. of the first minors; and so, as in § 7 of Chapter III, p. 25, $U(\lambda)$ is the corresponding invariant factor. In a similar manner it is easy to prove that the H.C.F. of second minors of $|B - \lambda I|$ is that particular minor which involves the $p + q - 2$ units belonging to the superdiagonals of the two highest submatrices $B_p(\lambda)$ and $B_q(\lambda)$, the first column and last row of each of these submatrices being thus deleted. Proceeding in this way we may verify that the invariant factors of the λ-matrix $A - \lambda I$, as defined in Chapter III, but here reversed, are $|B_p(\lambda)|$, $|B_q(\lambda)|$, . . . $|B_t(\lambda)|$, 1, 1, . . . , 1, or $U(\lambda)$, $V(\lambda)$, $W(\lambda)$, . . . , 1, 1, . . . , 1. (28)

The identification of the invariant factors of $B - \lambda I$ shows that $B - \lambda I$, and therefore the canonical form B, is unique for A and all collineatory transforms of A.

9. **Historical Note.**—A faint foreshadowing of a canonical matrix of the type B may be noticed in an observation of Spottiswoode (1853) that a binary quantic $a_0x^n - a_1x^{n-1}y + a_2x^{n-2}y^2 - \ldots + (-)^n a_n y^n$ could be expressed as a determinant of the type later called "recurrent", namely,

$$\begin{vmatrix} a_0 & a_1 & a_2 & \ldots & . & a_n \\ y & x & . & \ldots & . & . \\ . & y & x & \ldots & . & . \\ . & . & . & . & . & . \\ . & . & . & \ldots & y & x \end{vmatrix}.$$

(See Muir's *History*, II, p. 211.) With but little modification this appears as the characteristic determinant of a matrix of an elementary canonical form B. But Cayley's memoir on matrices had not then been written, and Spottiswoode obviously regarded the determinant as interesting but without application. Frobenius, to whom the theory of the reduced characteristic function of a matrix is due, in a long classical paper (*J. für Math.* **86**, (1879), 146–208, (206)) explicitly gives the rational canonical form B (as well as the classical irrational form which we shall next consider), as the solution of a problem equivalent to this: to construct a λ-matrix which shall have prescribed invariant factors. G. Landsberg (*J. für Math.* **116** (1896), 342) gives the bilinear form having matrix B_n as one of three specially simple canonical forms of an elementary bilinear form. A paper by W. Burnside (*Proc. Lond. Math. Soc., Ser.* I, **30** (1898), 183) is devoted to establishing the canonical form B in its full generality. In the present century the topic has been treated by Nicoletti, Lattès, Dickson, Kowalewski, Wedderburn, and others.

EXAMPLES

1. *Reduce to rational canonical form* $A = \begin{bmatrix} . & 1 & . & . \\ 2 & 3 & . & . \\ . & . & . & 1 \\ . & . & 3 & 2 \end{bmatrix}$.

This matrix *appears* to be in canonical form with $p = q = 2$. But closer inspection shows that the R.C.F.'s of the submatrices $\begin{bmatrix} . & 1 \\ 2 & 3 \end{bmatrix}$ and $\begin{bmatrix} . & 1 \\ 3 & 2 \end{bmatrix}$ have no common factor. This is an *elementary* matrix for which

$$\varphi = \psi = (A^2 - 3A - 2I)(A^2 - 2A - 3I) = A^4 - 5A^3 + A^2 + 13A + 6I.$$

The vectors (1, 0, 0, 0), (0, 0, 1, 0) are each of the second grade, but their *sum* (1, 0, 1, 0) is of fourth grade. Taking this to be the initial u, and applying Theorem II, we reduce A to

$$HAH^{-1} = \begin{bmatrix} . & 1 & . & . \\ . & . & 1 & . \\ . & . & . & 1 \\ -6 & -13 & -1 & 5 \end{bmatrix}.$$

2. Calculate H in Ex. 1, using $u = (1, 0, 1, 0)$, $H = \{u, uA, uA^2, uA^3\}$.

3. Show that, if A is non-singular, the vectors $u, uA, uA^2, \ldots$ have equal grades.

[$U(A)$ cannot have A as factor.]

4. Reduce $\begin{bmatrix} . & 1 & 1 \\ 1 & . & 1 \\ 1 & 1 & . \end{bmatrix}$, $\begin{bmatrix} 1 & 2 & 3 \\ 3 & 2 & 1 \\ 1 & 1 & 1 \end{bmatrix}$ to canonical forms.

5. Verify the identity $[x, y, z] \begin{bmatrix} c_1 & 1 & . \\ c_2 & . & 1 \\ c_3 & . & . \end{bmatrix} = [c_1x + c_2y + c_3z,\ x,\ y]$ where each of x, y, z is a column vector of order n.

Extend this to the case of p such vectors.

6. If $\psi(\lambda) = \lambda^p - c_1\lambda^{p-1} - \ldots - c_p$ is the R.C.F. of the matrix A_1 and x is a column vector of maximal grade p, show that $AH = HB_p$, where

$$H = [A^{p-1}x, A^{p-2}x, \ldots, Ax, x] \quad \text{and} \quad B_p = \begin{bmatrix} c_1 & 1 & . & \cdots & . \\ c_2 & . & 1 & \cdots & . \\ \hline c_{p-1} & . & . & \cdots & 1 \\ c_p & . & . & \cdots & . \end{bmatrix}.$$

7. Examine the rational canonical form of $A = \begin{bmatrix} . & 1 & . & . \\ 2 & 3 & q & . \\ . & . & . & 1 \\ . & . & 2 & 3 \end{bmatrix}$ (i) when $q = 0$ and (ii) when $q = 1$.

[(i) $\psi(A) = A^2 - 3A - 2I$, and A is already canonical and has two equal invariant factors: (ii) $\psi(A) = (A^2 - 3A - 2I)^2$, and A has one invariant factor. Maximal grades 2 and 4 respectively.]

CHAPTER VI

THE CLASSICAL CANONICAL FORM FOR THE COLLINEATORY GROUP

The canonical form B of the previous chapter, though obtained by rational operations in the field of A, was not the earliest to be discovered. The most important of its precursors was a form C which, in the case where the latent roots of A were all distinct, was a purely diagonal matrix having those latent roots, in any prescribed order, in the diagonal; so that $C = [\lambda_i \delta_{ij}]$. Such a form C is thus derivable in general only by irrational operations, since a necessary preliminary is the solution of the characteristic equation of A. When A has repeated roots it may well happen, as we shall find, that C cannot be purely diagonal; and perhaps the simplest way of ascertaining the requisite modifications will be to reverse the order of history, and deduce the classical form C from the rational canonical form B. We shall adopt this course first; later we shall give a second derivation, closer to the historical order of development, but at the same time more tentative.

1. The Classical Canonical Form deduced from the Rational Form.

We have $A = H_1 B H_1^{-1}$; if we can find H such that $B = HCH^{-1}$, then we shall have $A = H_2 C H_2^{-1}$, where $H_2 = H_1 H$. The point at issue is: what kind of matrix H transforms B into C? Let us suppose, for example, that $B = HCH^{-1}$, where

$$B = \begin{bmatrix} \cdot & 1 & \cdot & \cdot \\ \cdot & \cdot & 1 & \cdot \\ \cdot & \cdot & \cdot & 1 \\ b_4 & b_3 & b_2 & b_1 \end{bmatrix}, \qquad C = \begin{bmatrix} \alpha & \cdot & \cdot & \cdot \\ \cdot & \beta & \cdot & \cdot \\ \cdot & \cdot & \gamma & \cdot \\ \cdot & \cdot & \cdot & \delta \end{bmatrix}, \qquad (1)$$

and where the latent roots α, β, γ, δ are all distinct, and the characteristic function, common to A, B, and C, is given by

$$\begin{aligned} \phi(\lambda) = | A - \lambda I | &= \lambda^4 - b_1 \lambda^3 - b_2 \lambda^2 - b_3 \lambda - b_4 \\ &= (\lambda - \alpha)(\lambda - \beta)(\lambda - \gamma)(\lambda - \delta). \quad . \quad . \quad (2) \end{aligned}$$

If, as in § 4 of Chapter V, p. 49, we take an arbitrary vector $u = [u_1, u_2, u_3, u_4]$, where none of the components is zero, in order to generate the chain u, uC, uC^2, uC^3, and then take these four vectors to be the rows of H, we find at once that H is of the form

$$H = \begin{bmatrix} u_1 & u_2 & u_3 & u_4 \\ u_1\alpha & u_2\beta & u_3\gamma & u_4\delta \\ u_1\alpha^2 & u_2\beta^2 & u_3\gamma^2 & u_4\delta^2 \\ u_1\alpha^3 & u_2\beta^3 & u_3\gamma^3 & u_4\delta^3 \end{bmatrix}. \quad . \quad . \quad . \quad (3)$$

By a familiar result in determinants H is non-singular, since

$$\begin{aligned} |H| &= u_1u_2u_3u_4 \,|\, \alpha^0\beta^1\gamma^2\delta^3 \,| \\ &= u_1u_2u_3u_4\,\Delta(\alpha\beta\gamma\delta), \quad . \quad . \quad . \quad . \quad . \quad (4) \end{aligned}$$

where $\Delta(\alpha\beta\gamma\delta)$ denotes the difference-product

$$(\delta-\gamma)(\delta-\beta)(\delta-\alpha)(\gamma-\beta)(\gamma-\alpha)(\beta-\alpha).$$

If for simplicity we put $u = [1, 1, 1, 1]$, we have at once that $B = HCH^{-1}$, or $BH = HC$, where the last relation, written out in full, appears as

$$\begin{bmatrix} . & 1 & . & . \\ . & . & 1 & . \\ . & . & . & 1 \\ b_4 & b_3 & b_2 & b_1 \end{bmatrix} \begin{bmatrix} 1 & 1 & 1 & 1 \\ \alpha & \beta & \gamma & \delta \\ \alpha^2 & \beta^2 & \gamma^2 & \delta^2 \\ \alpha^3 & \beta^3 & \gamma^3 & \delta^3 \end{bmatrix}$$

$$= \begin{bmatrix} 1 & 1 & 1 & 1 \\ \alpha & \beta & \gamma & \delta \\ \alpha^2 & \beta^2 & \gamma^2 & \delta^2 \\ \alpha^3 & \beta^3 & \gamma^3 & \delta^3 \end{bmatrix} \begin{bmatrix} \alpha & . & . & . \\ . & \beta & . & . \\ . & . & \gamma & . \\ . & . & . & \delta \end{bmatrix} = \begin{bmatrix} \alpha & \beta & \gamma & \delta \\ \alpha^2 & \beta^2 & \gamma^2 & \delta^2 \\ \alpha^3 & \beta^3 & \gamma^3 & \delta^3 \\ \alpha^4 & \beta^4 & \gamma^4 & \delta^4 \end{bmatrix}. \quad (5)$$

This may easily be verified by means of (2), and it shows incidentally that the product matrix BH, or HC, is of a type similar to H. Since the name "alternant" is now well established for the determinant of a matrix of this kind, we may perhaps call H and BH *alternant* matrices. For this particular case, then, of canonical matrices B and C, it appears that B and C are each quotients of the same two alternant matrices, B being obtained by postdivision, C by predivision. The result may be extended at once to the case of a matrix A with its n latent roots all different.

Considering still the elementary case of B, let us next suppose that certain of the latent roots are multiple. The alternant matrix H then becomes singular, since two or more columns are identical. In the theory of determinants, however, a modified alternant is met with in this case: namely, if a latent root a is m-fold, and the columns of the alternant containing it are numbered $0, 1, 2, \ldots, m-1$, then the column 0 is the same as before, but the elements of the column j are derived from those of column 0 by the operation $\left(\frac{d}{da}\right)^j/j!$. For example, if the latent roots are $a, a, a, \beta, \beta, \gamma$, the modified matrix H appears as

$$H = \begin{bmatrix} 1 & . & . & 1 & . & 1 \\ a & 1 & . & \beta & 1 & \gamma \\ a^2 & 2a & 1 & \beta^2 & 2\beta & \gamma^2 \\ a^3 & 3a^2 & 3a & \beta^3 & 3\beta^2 & \gamma^3 \\ a^4 & 4a^3 & 6a^2 & \beta^4 & 4\beta^3 & \gamma^4 \\ a^5 & 5a^4 & 10a^3 & \beta^5 & 5\beta^4 & \gamma^5 \end{bmatrix} \quad a \neq \beta,\ \beta \neq \gamma,\ a \neq \gamma, \quad (6)$$

while the determinant takes the value

$$|H| = \Delta\{(aaa)(\beta\beta)\gamma\} = (\gamma-\beta)^2(\gamma-a)^3(\beta-a)^6, \quad (7)$$

which is a generalized or *confluent* difference-product, so that H is non-singular. If we set up again the equation $BH = HC$, provisionally placing the latent roots in the diagonal of C as before, with zeros below, but leaving the elements above the diagonal to be determined by inspecting the products, we find that C differs from the previous case only in having *a unit in the superdiagonal at every position adjacent to a pair of repeated roots.* We have in fact

$$\begin{bmatrix} . & 1 & . & . & . & . \\ . & . & 1 & . & . & . \\ . & . & . & 1 & . & . \\ . & . & . & . & 1 & . \\ . & . & . & . & . & 1 \\ b_6 & b_5 & b_4 & b_3 & b_2 & b_1 \end{bmatrix} \begin{bmatrix} 1 & . & . & 1 & . & 1 \\ a & 1 & . & \beta & 1 & \gamma \\ a^2 & 2a & 1 & \beta^2 & 2\beta & \gamma^2 \\ a^3 & 3a^2 & 3a & \beta^3 & 3\beta^2 & \gamma^3 \\ a^4 & 4a^3 & 6a^2 & \beta^4 & 4\beta^3 & \gamma^4 \\ a^5 & 5a^4 & 10a^3 & \beta^5 & 5\beta^4 & \gamma^5 \end{bmatrix}$$

$$= \begin{bmatrix} 1 & . & . & 1 & . & 1 \\ a & 1 & . & \beta & 1 & \gamma \\ a^2 & 2a & 1 & \beta^2 & 2\beta & \gamma^2 \\ a^3 & 3a^2 & 3a & \beta^3 & 3\beta^2 & \gamma^3 \\ a^4 & 4a^3 & 6a^2 & \beta^4 & 4\beta^3 & \gamma^4 \\ a^5 & 5a^4 & 10a^3 & \beta^5 & 5\beta^4 & \gamma^5 \end{bmatrix} \begin{bmatrix} a & 1 & . & . & . & . \\ . & a & 1 & . & . & . \\ . & . & a & . & . & . \\ . & . & . & \beta & 1 & . \\ . & . & . & . & \beta & . \\ . & . & . & . & . & \gamma \end{bmatrix} =$$

$$\begin{bmatrix} \alpha & 1 & \cdot & \beta & 1 & \gamma \\ \alpha^2 & 2\alpha & 1 & \beta^2 & 2\beta & \gamma^2 \\ \alpha^3 & 3\alpha^2 & 3\alpha & \beta^3 & 3\beta^2 & \gamma^3 \\ \alpha^4 & 4\alpha^3 & 6\alpha^2 & \beta^4 & 4\beta^3 & \gamma^4 \\ \alpha^5 & 5\alpha^4 & 10\alpha^3 & \beta^5 & 5\beta^4 & \gamma^5 \\ \alpha^6 & 6\alpha^5 & 15\alpha^4 & \beta^6 & 6\beta^5 & \gamma^6 \end{bmatrix}, \quad . \quad . \quad . \quad . \quad . \quad (8)$$

the identity of the elements in the extended products depending on the simplest additive properties of binomial coefficients, together with the following appropriate conditions satisfied by the repeated and unrepeated roots,

$$\phi(\lambda) = \lambda^6 - b_1\lambda^5 - b_2\lambda^4 - b_3\lambda^3 - b_4\lambda^2 - b_5\lambda - b_6 = 0, \quad \lambda = \alpha, \beta, \gamma,$$

$$\phi'(\lambda) = 6\lambda^5 - 5b_1\lambda^4 - 4b_2\lambda^3 - 3b_3\lambda^2 - 2b_4\lambda - b_5 = 0, \quad \lambda = \alpha, \beta,$$

$$\frac{1}{2!}\phi''(\lambda) = 15\lambda^4 - 10b_1\lambda^3 - 6b_2\lambda^2 - 3b_3\lambda - b_4 = 0, \qquad \lambda = \alpha. \qquad (9)$$

This transformation of an elementary canonical matrix B_k by an alternant matrix is readily seen to be of general application; the diagonally placed submatrices in the derived classical form are of typical appearance

$$C_1(\alpha) = \alpha, \quad C_2(\alpha) = \begin{bmatrix} \alpha & 1 \\ \cdot & \alpha \end{bmatrix}, \quad C_3(\alpha) = \begin{bmatrix} \alpha & 1 & \cdot \\ \cdot & \alpha & 1 \\ \cdot & \cdot & \alpha \end{bmatrix}, \quad \ldots, \qquad (10)$$

and there is one of these isolated submatrices for each distinct latent root of B_k. We may perhaps call them *simple classical submatrices.* Since each elementary submatrix B_k of the rational canonical form B is isolated it can be transformed independently of the rest, so that *the classical canonical form C is finally seen to consist of simple classical submatrices $C_k(\lambda)$, isolated along the diagonal, all other elements being zero.* It is to be observed that the same latent root α may be associated with several classical submatrices, the sum of the orders of these being the total multiplicity of α.

We shall write this form as

$$C = [C_k(\lambda)], \quad . \quad . \quad . \quad . \quad . \quad . \quad . \quad . \quad (11)$$

where $k = k_1, k_2, \ldots, k_\mu$; $\lambda = \alpha, \beta, \ldots$. In (8) above, we have $\mu = 3$ and $C_k(\lambda) = C_3(\alpha), C_2(\beta), C_1(\gamma)$. Using a convenient notation called by Segre the *characteristic* of the elementary divisors, we collect together the various suffixes k and write [321] to characterize C. If, however, C were to consist of $C_3(\alpha)$, $C_2(\alpha)$, $C_1(\gamma)$, $\alpha \neq \gamma$, we should collect together the suffixes belonging to α and write [(32)1]. The characteristic of an *elementary* matrix B_k accordingly differs from that of a more elaborate matrix B in having no enclosing round-brackets ( ).

2. The Auxiliary Unit Matrix.

The typical simple classical submatrix $C_k(\lambda)$ may be written $\lambda I + U$, where U is a matrix with units in the superdiagonal and zeros everywhere else. Thus, for the fourth order,

$$U = \begin{bmatrix} \cdot & 1 & \cdot & \cdot \\ \cdot & \cdot & 1 & \cdot \\ \cdot & \cdot & \cdot & 1 \\ \cdot & \cdot & \cdot & \cdot \end{bmatrix},$$

and in general

$$U = [\epsilon_{ij}], \text{ where } \epsilon_{ij} \begin{cases} = 0, & i \neq j - 1, \\ = 1, & i = j - 1. \end{cases} \qquad . \quad . \quad (12)$$

This matrix U, which has already played a part in the rational elementary submatrices B_k, is so useful an instrument in the algebra of matrices that it seems worthy of a special name. We shall call it the *auxiliary unit matrix*. The following are some of its simpler properties:

I. The matrices U, U^2, U^3, ..., U^{p-1}, where U is of order p, each consist of units filling a single diagonal, respectively the first, second, ..., $(p-1)$th superdiagonal. The transposed matrices $(U')^k$ similarly occupy the subdiagonals.

II. The Reduced Characteristic Equation of U is $U^p = 0$, and higher powers of U are zero. The rank of U^k is $p - k$, for $k \leqslant p$.

III. If $C_k(\lambda) \equiv C_k = \lambda I + U$, any polynomial function of C_k, in virtue of the commutative property of I, can be expressed as a Taylor's series in powers of U, thus:

$$\begin{aligned} f(C_k) &= f(\lambda I + U) \\ &= f(\lambda) I + f'(\lambda) U + \frac{1}{2!} f''(\lambda) U^2 + \ldots + \frac{1}{(k-1)!} f^{(k-1)}(\lambda) U^{k-1}, \end{aligned}$$

Hence the matrix $f(C_k)$ can at once be written down. For example,

$$\text{if} \quad C_k = \begin{bmatrix} \lambda & 1 & . & . \\ . & \lambda & 1 & . \\ . & . & \lambda & 1 \\ . & . & . & \lambda \end{bmatrix}, \quad \text{then } C_k^3 = \begin{bmatrix} \lambda^3 & 3\lambda^2 & 3\lambda & 1 \\ . & \lambda^3 & 3\lambda^2 & 3\lambda \\ . & . & \lambda^3 & 3\lambda^2 \\ . & . & . & \lambda^3 \end{bmatrix},$$

and in general $f(C_k)$ is persymmetric.

EXAMPLES

1. If $U = \begin{bmatrix} . & 1 & . & . \\ . & . & 1 & . \\ . & . & . & 1 \\ . & . & . & . \end{bmatrix}$, then $U^2 = \begin{bmatrix} . & . & 1 & . \\ . & . & . & 1 \\ . & . & . & . \\ . & . & . & . \end{bmatrix}$, $U^3 = \begin{bmatrix} . & . & . & 1 \\ . & . & . & . \\ . & . & . & . \\ . & . & . & . \end{bmatrix}$, $U^4 = 0$.

2. Evaluate the integral powers of U'. If $X = U'U$ or UU', then $X^2 = X$. If $r + s$ is the order of U, show that $U^r(U')^r + (U')^s U^s = I$.

3. To prove that the alternant $|\alpha^0 \beta^1 \gamma^2 \delta^3|$ of (4) above is equal to the difference-product $\Delta(\alpha\beta\gamma\delta)$, we merely observe that it vanishes, having pairs of columns identical, when $\alpha = \beta$, $\alpha = \gamma$, &c. The numerical factor is fixed by inspecting the diagonal.

4. If instead of $\alpha^3, \beta^3, \ldots$, in the last row of the alternant we have $f(\alpha)$, $f(\beta), \ldots$, where $f(x)$ is a polynomial, we deduce similarly by the remainder theorem that the alternant $|\alpha^0 \beta^1 \gamma^2 f(\delta)|$ is divisible by the difference-product;

$$|\alpha^0 \beta^1 \gamma^2 f(\delta)| \equiv 0, \quad \operatorname{mod} \Delta(\alpha\beta\gamma\delta).$$

5. If we divide the alternant $|\alpha^0(\alpha+h)^1(\alpha+2h)^2\beta^3(\beta+h)^4\gamma^5|$ by $\Delta(0, h, 2h).\Delta(0, h)$ and proceed to the limit, $h \to 0$, we obtain the confluent alternant of (6) above. We also evaluate it, by means of Ex. 3, as the confluent difference-product in (7), namely $\Delta\{(\alpha\alpha\alpha)(\beta\beta)\gamma\}$.

In the general case, if α_i occurs in ν_i columns of a differentiated alternant, and α_j in ν_j columns, where $\alpha_i \neq \alpha_j$, $i < j$, then the alternant is equal in value to the confluent difference-product $\prod_{ij} (\alpha_j - \alpha_i)^{\nu_i \nu_j}$.

6. If in the last row of the first alternant in Ex. 5 we write $f(\alpha)$, &c., instead of $\alpha^5, \ldots$, and proceed as directed, we obtain for the last row of the resulting alternant the elements $f(\alpha)$, $f'(\alpha)$, $f''(\alpha)/2!$, $f(\beta)$, $f'(\beta)$, $f(\gamma)$.

We deduce also, just as in Ex. 4, that if $f(x)$ is a polynomial the general differentiated alternant is divisible by the confluent difference-product; e.g.

$$|\alpha^0\alpha^0\alpha^0\beta^3\beta^3 f(\gamma)| \equiv 0, \quad \operatorname{mod} \Delta\{(\alpha\alpha\alpha)(\beta\beta)\gamma\}.$$

7. Find the rational canonical form B corresponding to the classical form

$$C = \begin{bmatrix} \alpha & 1 & . & . & . \\ . & \alpha & . & . & . \\ . & . & \alpha & . & . \\ . & . & . & \beta & 1 \\ . & . & . & . & \beta \end{bmatrix}.$$

[If $\alpha \neq \beta$, the leading submatrix is B_4, the corresponding invariant factor being the quartic polynomial $(\lambda - \alpha)^2 (\lambda - \beta)^2$. Thus B has two submatrices B_4, $B_1 = [\alpha]$ and two invariant factors. Characteristic [(21) 2].

If $\alpha = \beta$, then B has three submatrices B_2, B_2, B_1 and three invariant factors $(\lambda - \alpha)^2$, $(\lambda - \alpha)^2$, $(\lambda - \alpha)$. Characteristic [(221)].

8. The classical form corresponding to

$$B = \begin{bmatrix} \cdot & 1 & \cdot & \cdot \\ \cdot & \cdot & 1 & \cdot \\ \cdot & \cdot & \cdot & 1 \\ -4 & \cdot & 4 & \cdot \end{bmatrix} \text{ is } C = \begin{bmatrix} \sqrt{2} & 1 & \cdot & \cdot \\ \cdot & \sqrt{2} & \cdot & \cdot \\ \cdot & \cdot & -\sqrt{2} & 1 \\ \cdot & \cdot & \cdot & -\sqrt{2} \end{bmatrix}.$$

The characteristic of C is [22].

9. The number of Segre characteristics of order n without round brackets is $p(n)$, the number of partitions of the integer n. The number of ways of inserting round brackets in a characteristic of m indices is $p(m)$. Hence the number of canonical types for a given order can be found by combinatory analysis.

3. The Canonical Form of Jacobi.

The classical form C can also be obtained without the intermediate use of the rational form B, by a systematic but in general irrational reduction, which uses as a basis a canonical form due to Jacobi. Jacobi's form, like the classical, has the latent roots in the diagonal and zeros below, but the elements above the diagonal are not further specialized. Thus the form, Γ say, has the triangular appearance

$$\Gamma = \begin{bmatrix} \gamma_1 & a_{12} & a_{13} & a_{14} & a_{15} \\ \cdot & \gamma_2 & a_{23} & a_{24} & a_{25} \\ \cdot & \cdot & \gamma_3 & a_{34} & a_{35} \\ \cdot & \cdot & \cdot & \gamma_4 & a_{45} \\ \cdot & \cdot & \cdot & \cdot & \gamma_5 \end{bmatrix}. \quad . \quad . \quad . \quad (13)$$

The transformations used by Jacobi were of the equivalent kind, $\Gamma = PAQ$. We shall employ only collineatory operations.

Lemma.—*By a collineatory transformation the matrix* A *can be brought to the form* $\Gamma_1 = [\gamma_{ij}]$, *where the* n — 1 *subdiagonal elements* $\gamma_{21}, \gamma_{31}, \ldots, \gamma_{n1}$ *in the first column are zero.*

Proof.—Since every square matrix has at least one pole (Chapter IV, p. 41) corresponding to a latent root λ_i, the matrix A has a pole $y = \{y_1, y_2, \ldots, y_n\}$, where at least one component y_k is non-zero: thus $Ay = \lambda_i y$. It will be convenient to take y_k as equal to unity, which is possible since the co-ordinates are homogeneous. If the pole

y were the point $\xi = \{1, 0, 0, \ldots, 0\}$, then A would already be in the form Γ_1, since we may verify at once that $\Gamma_1\xi = \gamma_{11}\xi$.

If, however, this is not the case, we construct a non-singular matrix H_1, such that $H_1 y = \xi$; for example, if $n = 5$, $k = 3$, the matrix H_1 will have the appearance

$$H_1 = \begin{bmatrix} \cdot & \cdot & 1 & \cdot & \cdot \\ \cdot & \cdot & y_4 & -1 & \cdot \\ \cdot & \cdot & y_5 & \cdot & -1 \\ -1 & \cdot & y_1 & \cdot & \cdot \\ \cdot & -1 & y_2 & \cdot & \cdot \end{bmatrix}, \quad |H_1| \neq 0. \qquad (14)$$

The kth column contains $y_k, y_{k+1}, \ldots, y_n, y_1, \ldots, y_{k-1}$ in this order, and what may be termed *the* kth *complete diagonal* is filled up with elements $-y_k$, where $y_k = 1$. By multiplication it may be verified that $H_1 y = \xi$. Since also $Ay = \lambda_i y$, we have at once $H_1 A H_1^{-1}\xi = \lambda_i \xi$.

Thus $H_1 A H_1^{-1}$ is of the form Γ_1, where $\gamma_{11} = \lambda_i$, and the Lemma is proved. It is to be noted that λ_i may quite well be zero.

Theorem.—*For any given square matrix* $\mathrm{A} = [\mathrm{a}_{ij}]$ *there exists a non-singular matrix* H, *which transforms* A *into a Jacobian canonical form* Γ *such that* $\Gamma = \mathrm{HAH}^{-1}$.

Proof.—The result follows by repeated application of the Lemma. In the first place the matrix A is transformed by H_1 to $\Gamma_1 = H_1 A H_1^{-1}$. If A_1 is the submatrix obtained by deleting the first row and first column of Γ_1, then A_1 can again, by the Lemma, be reduced to a similar form $\Gamma_2 = H_2 A_1 H_2^{-1}$, so that the transform of A,

$$\begin{bmatrix} 1 & \cdot \\ \cdot & H_2 \end{bmatrix} \Gamma_1 \begin{bmatrix} 1 & \cdot \\ \cdot & H_2^{-1} \end{bmatrix} = \begin{bmatrix} 1 & \cdot \\ \cdot & H_2 \end{bmatrix} H_1 A H_1^{-1} \begin{bmatrix} 1 & \cdot \\ \cdot & H_2 \end{bmatrix}^{-1} \qquad (15)$$

now has subdiagonal zeros in the *first two* columns. By at most $n - 1$ such steps, each involving collineatory operations on submatrices of lower and lower order, we replace all subdiagonal elements by zero. (The superdiagonal elements do not at present concern us.) The continued product of the transforming matrices $\begin{bmatrix} 1 & \cdot \\ \cdot & H_k \end{bmatrix}$ thus yields a matrix H, which reduces A to Jacobian form $HAH^{-1} = \Gamma$.

Corollary.—*In* Γ *the diagonal elements are the* n *latent roots of* A, *arranged in any prescribed order*: for $| A - \lambda I | = | \Gamma - \lambda I |$, which on expansion is obviously equal to $(-)^n (\lambda - \gamma_1)(\lambda - \gamma_2) \ldots (\lambda - \gamma_n)$.

Also the steps of the proof show that the latent points, and therefore the latent roots, can be taken in any order.

It is usual to arrange the roots according to the frequency of repeated roots, in descending order of multiplicity, the convention being displayed by the example $\alpha, \alpha, \alpha, \beta, \beta, \gamma$. The diagonal array of latent roots in a canonical form C or Γ forms, as it were, the vertebra of the matrix; perhaps it could be called the *nucleus* of A or of any transform HAH^{-1}.

4. The Classical Canonical Form deduced from that of Jacobi.

It will next be shown how, by collineatory operations of Types I, II, and III (p. 11), the elements above the diagonal in the Jacobian canonical form Γ may be eliminated or adjusted in such a way as to yield the classical form C. A guiding principle in the later stages is the invariant property of *chains* formed by linking non-diagonal elements in a manner now to be described. Consider, for example, the following combination of two canonical submatrices corresponding to the same latent root α, namely

$$\begin{bmatrix} C_4(\alpha) & \\ & C_3(\alpha) \end{bmatrix} \equiv \begin{bmatrix} \alpha & 1 & . & . & & & \\ & \alpha & 1 & . & & & \\ & & \alpha & 1 & & & \\ & & & \alpha & & & \\ & & & & \alpha & 1 & . \\ & & & & & \alpha & 1 \\ & & & & & & \alpha \end{bmatrix}. \qquad (16)$$

The non-zero elements above the diagonal occupy places which fall into two chains (1, 2), (2, 3), (3, 4) and (5, 6), (6, 7), the element (1, 2), for example, being linked to (2, 3) by the fact that the column-index of the former is the row-index of the latter. The two chains are *isolated*, in that they have no row-index or column-index in common; and an isolated chain of length $m - 1$ involves an isolated submatrix of order m. It is easy to see that collineatory substitutions of Type I (p. 11) leave the properties of chain-length and isolation invariant; for example, the chains (1, 3), (3, 5), (5, 7) and (2, 4), (4, 6) agree in these respects with the two we have given above, and may be transformed into them. The steps are the interchanges I_{ij}, where $ij = 23$, 45, 34, 67, 56, 45 in this succession.

It is a help to form staircase graphs of these chains as follows:

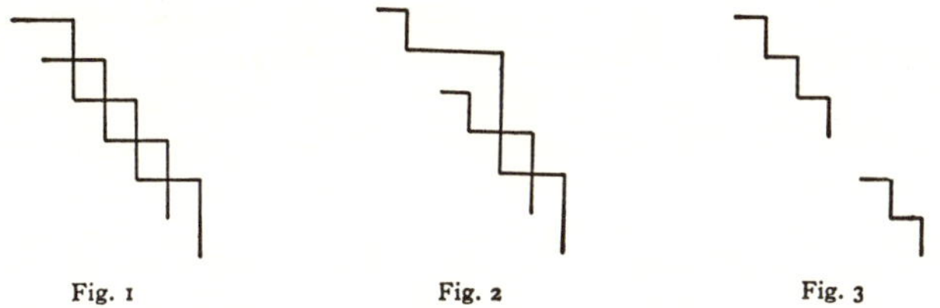

Fig. 1 Fig. 2 Fig. 3

Fig. 1 indicates these chains before interchange: fig. 2 after the first interchange I_{23}: fig. 3 after the final interchange I_{45}. While geometrical length of a chain varies, chain-length—the number of steps—is fixed. The chains being isolated, the zigzags are never collinear. What the interchange process does is systematically to disentangle the chains.

We consider then the Jacobian canonical form Γ. Suppose first that there is a certain non-zero element a_{ij} above the diagonal, where the latent roots aligned with it, λ_i and λ_j, are distinct. The collineatory operation $\text{col}_j - h\ \text{col}_i$, $\text{row}_i + h\ \text{row}_j$ replaces a_{ij} by $a_{ij} - h(\lambda_i - \lambda_j)$, and in general also modifies elements in the ith row to the *right* of a_{ij}, and in the jth column *above* a_{ij}, as may readily be verified. If we take $h = a_{ij}/(\lambda_i - \lambda_j)$, as we may, since $\lambda_i \neq \lambda_j$, then a_{ij} is replaced by zero. Now the elements such as a_{ij} which are aligned with distinct latent roots lie in rectangles; and by operations like the above we can clear these row by row, beginning at the left of the lowest row. When this is done the submatrices of Jacobian shape corresponding to single latent roots are isolated, and may therefore be considered separately.

We consider then a submatrix Γ_m of order m, in a form such as

$$\Gamma_5 = \begin{bmatrix} a & a_{12} & a_{13} & a_{14} & a_{15} \\ . & a & a_{23} & a_{24} & a_{25} \\ . & . & a & a_{34} & a_{35} \\ . & . & . & a & a_{45} \\ . & . & . & . & a \end{bmatrix} . \quad . \quad . \quad . \quad (17)$$

If the non-zero elements above the diagonal are such that there is *not more than one in any row or column*, we may link them in chains, and, by bringing them into superdiagonal position by operations of Type I, and transforming them into units by operations of Type III, as in Chapter II, § 4, p. 13, obtain simple classical submatrices.

If, however, this is not the case we proceed by induction, the induction resting on the initial fact that for the case $m = 2$ the matrix Γ_2 can at once be transformed into classical shape by an operation of Type III. Let us take Γ_m in the form

$$\Gamma_m = \begin{bmatrix} \Gamma_{m-1} & A_1 \\ \cdot & a \end{bmatrix}, \qquad \ldots \ldots \quad (18)$$

where Γ_{m-1} denotes the leading submatrix of order $m-1$, and A_1 denotes a column matrix of $m-1$ elements; and let us assume that Γ_{m-1} can be transformed into classical shape C_{m-1} by a matrix H_1.

Then we must have

$$\begin{bmatrix} H_1 & \cdot \\ \cdot & I \end{bmatrix} \begin{bmatrix} \Gamma_{m-1} & A_1 \\ \cdot & a \end{bmatrix} \begin{bmatrix} H_1^{-1} & \cdot \\ \cdot & I \end{bmatrix} = \begin{bmatrix} C_{m-1} & D_1 \\ \cdot & a \end{bmatrix}, \quad (19)$$

where $D_1 = H_1A_1$ and where, written out in full, the right-hand matrix has the appearance, e.g.,

$$\begin{bmatrix} a & 1 & \cdot & \cdot & \cdot & \cdot & d_{17} \\ \cdot & a & 1 & \cdot & \cdot & \cdot & d_{27} \\ \cdot & \cdot & a & \cdot & \cdot & \cdot & d_{37} \\ \cdot & \cdot & \cdot & a & 1 & \cdot & d_{47} \\ \cdot & \cdot & \cdot & \cdot & a & 1 & d_{57} \\ \cdot & \cdot & \cdot & \cdot & \cdot & a & d_{67} \\ \cdot & \cdot & \cdot & \cdot & \cdot & \cdot & a \end{bmatrix}, \qquad m = 7. \quad . \ (20)$$

If any element d_{km} has a superdiagonal unit in the same row, the operation $\text{col}_m - d_{km}\text{col}_{k+1}$, $\text{row}_{k+1} + d_{km}\text{row}_m$ simply removes d_{km} without further change. Hence we may remove any such elements d_{km}, so that now there is at most one non-zero element, excluding the diagonal, in each row. For instance, in (20) we may at once take $d_{17} = d_{27} = d_{47} = d_{57} = 0$. If there still remains more than one non-zero element in the mth column (as here if $d_{37} \neq 0$, $d_{67} \neq 0$), we form chains and proceed in a manner exemplified without loss of generality by the chains (1, 2), (2, 3), (3, 7) and (4, 5), (5, 6), (6, 7). We shall show that the element (6, 7), or d_{67}, can be annulled in three moves by the earlier chain. In fact the operation

$\text{row}_6 - h\ \text{row}_3$, $\text{col}_3 + h\ \text{col}_6$, where $h = d_{67}/d_{37}$, annuls (6, 7), but introduces a (5, 3) because of the (5, 6). This (5, 3) is similarly annulled by means of (2, 3), but a (4, 2) is introduced through the (4, 5). Lastly the (4, 2) is annulled by the (1, 2), and no element is introduced, because the chain commencing with (4, 5) gives out, having no earlier element. It thus appears in general that all elements in the last column may be removed by collineatory operations *except the one which belongs to the chain of greatest length*, an earlier chain being preferred to a later in the case of alternative chains of equal length. Lastly the isolated chains may be brought into classical superdiagonal position by operations of Type I, and if the element brought there from the last column is not unity, it can be made so by an operation of Type III.

The induction is therefore complete and, when carried out step by step and applied to the separate submatrices, it provides a means of bringing a matrix in Jacobian form Γ to classical form C by collineatory transformations.

5. Uniqueness of the Classical Form: Elementary Divisors.

It was found in dealing with the rational canonical form B that the invariant factors of the characteristic matrix $A - \lambda I$ were unique, being given by the determinants $|B_k(\lambda)|$. The distribution of latent roots among these determinants is therefore invariant, and thus a factor like $(\lambda - \lambda_j)$ will occur in the successive determinants $|B_k(\lambda)|$ with invariant exponents $e_{1j}, e_{2j}, \ldots$, where, since each $|B_k(\lambda)|$ is, as we have seen, a factor of the preceding one, we must necessarily have $e_{1j} \geqslant e_{2j} \geqslant \ldots$.

The factors $(\lambda - \lambda_j)^{e_{ij}}$ are called the *elementary divisors corresponding to the latent root λ_j of the matrix A.* From the manner in which the form C has been deduced from the form B, we see at once that *these elementary divisors are the characteristic determinants of the simple classical submatrices in the canonical form C.* This amounts to identifying the exponents of the elementary divisors with the orders of the simple classical submatrices, and could alternatively be established by investigating the H.C.F. of minors of various orders in the characteristic determinant of C, just as was done for B in Chapter V, p. 55, the superdiagonal units playing an analogous rôle. The elementary divisors are invariants in a field which includes the latent roots as well as the elements of A; and they are related to the invariant factors $E_i(\lambda)$ thus (with the order of p. 23 reversed):

$$E_i(\lambda) = (\lambda - \lambda_1)^{e_{i1}} (\lambda - \lambda_2)^{e_{i2}} (\lambda - \lambda_3)^{e_{i3}} \ldots, \quad i = 1, 2, \ldots \quad (21)$$

The Segre characteristic $[(e_{11}e_{21}\ldots)(e_{12}e_{22}\ldots)\ldots]$ is now more naturally written as the integer matrix

$$[e_{ij}] = \begin{bmatrix} e_{11} & e_{12} & \cdots \\ e_{21} & \cdots & \cdot \\ \cdots & \cdots & \cdot \end{bmatrix}, \quad \sum_i \sum_j e_{ij} = n, \qquad (22)$$

where $\sum_j e_{1j} = p$, $\sum_j e_{2j} = q, \ldots$; and where $e_{1j} \geqslant e_{2j} \geqslant \ldots$.

The sum k of the integers on the ith row gives the order of the ith invariant factor $|B_k(\lambda)|$. The sum k' of the jth column gives the degree of repetition of the jth distinct latent root λ_j.

From these facts we see that the canonical form C is unique, and that collineatory equivalence, or similarity, of matrices is completely expressed by the invariance of the elementary divisors, or (if we prefer) of the invariant factors. Further points of interest are brought out in the following examples.

EXAMPLES

1. The elementary divisors of the matrix C in Ex. 7, § 2, p. 63, are $(\lambda - \alpha)^2$, $(\lambda - \alpha)$, $(\lambda - \beta)^2$. The characteristic is $\begin{bmatrix} 2 & 2 \\ 1 & \end{bmatrix}$.

2. Reduce to the canonical forms B, C, and Γ the matrix

$$A = \begin{bmatrix} 2 & . & 2 & 1 \\ 6 & 1 & 4 & 4 \\ 10 & . & . & 4 \\ 7 & . & -7 & 2 \end{bmatrix},$$

having given that one unrepeated latent root is 2, and that the others involve repetitions.

3. *Derive relation* (8) *directly from the theorem on p.* 49 *by finding the rational canonical form of the classical.*

Let the matrix A of p. 49, whose rational canonical form is B, be taken to be $A = [C_k(\lambda)]$, where $C_k(\lambda) = C_3(\alpha)$, $C_2(\beta)$, $C_1(\gamma)$, and where α, β, γ all differ. Starting with the vector $u = [1, 0, 0, 1, 0, 1]$ construct the chain u, uA, uA^2, uA^3, uA^4, uA^5. These six vectors will be found to give the required H, namely, the second matrix of relation (8) above.

The computation of this chain illustrates property III of § 2. Any vector u can be chosen, provided that its first, fourth, and sixth elements are non-zero. We have naturally chosen the simplest, u, consistent with the condition $|H| \neq 0$. This method applies to any A.

4. To derive the Classical Form by the Methods of Chapter V.

First reduce A to the rational B, finding its R.C.F. $\varphi(A)$ which must now be factorized.

Let $\varphi(\lambda) = (\lambda - \alpha)^{p_1}(\lambda - \beta)^{p_2}\dots(\lambda - \gamma)^{p_\nu} = (\lambda - \alpha)^{p_1}\psi(\lambda)$,

where $\alpha, \beta, \dots, \gamma$ are the ν distinct latent roots: $\psi(\lambda)$ denotes all the factors not involving α, so that $\psi(\alpha) \neq 0$. Take an arbitrary vector u of maximum grade $p = (p_1 + p_2 + \dots + p_\nu)$: and now consider the vector

$$v = u\psi(A) = u(A - \beta I)^{p_2}\dots(A - \gamma I)^{p_\nu}.$$

This v cannot be identically zero since the R.C.F. of u is of degree greater by p_1 than the degree of ψ. The chain of vectors

$$v,\ v(A - \alpha I),\ v(A - \alpha I)^2,\ \dots,\ v(A - \alpha I)^{p_1 - 1}$$

terminates; for the next $v(A - \alpha I)^{p_1} = u\varphi(A) = 0$, since $\varphi(A) = 0$.

With the notation of p. 49, let $y = Ax$, $\xi = Hx$, $\eta = Hy$; also let

$$\begin{aligned} \xi_1 &= vx, \\ \xi_2 &= v(A - \alpha I)x, \\ &\dots\dots\dots \\ \xi_{p_1} &= v(A - \alpha I)^{p_1 - 1}x, \end{aligned}$$

so that the chain of vectors gives us the first p_1 rows of the matrix H. On expanding the first p_1 rows of $\eta = Hy$, we have

$$\begin{aligned} \eta_1 &= vy = vAx = \alpha\xi_1 + \xi_2, \\ \eta_2 &= v(A - \alpha I)y = v(A - \alpha I)Ax = \alpha\xi_2 + \xi_3, \\ &\dots\dots\dots\dots\dots\dots\dots\dots \\ \eta_{p_1} &= v(A - \alpha I)^{p_1 - 1}y = v(A - \alpha I)^{p_1 - 1}Ax = \alpha\xi_{p_1}, \end{aligned}$$

since $v(A - \alpha I)^{p_1 - 1}(A - \alpha I) = 0$. But $\eta = HAH^{-1}\xi$: hence the first p_1 rows of the classical canonical HAH^{-1} have been found, agreeing with the form already adopted, e.g. if $p_1 = 3$,

$$\begin{bmatrix} \eta_1 \\ \eta_2 \\ \eta_3 \end{bmatrix} = \begin{bmatrix} \alpha & 1 & . \\ . & \alpha & 1 \\ . & . & \alpha \end{bmatrix} \begin{bmatrix} \xi_1 \\ \xi_2 \\ \xi_3 \end{bmatrix}.$$

Further p_2 rows are derived from a chain initiated by the vector $u\{\varphi(A)/(A - \beta I)^{p_2}\}$: and so on.

An argument analogous to that of p. 51 justifies the linear independence of the first p rows of H.

5. *Given the characteristic* [(3, 2) (4, 1, 1) 5, 2] *and the distinct latent roots of a matrix* A, *determine its order, rank, invariant factors, elementary divisors, and its rational and classical canonical forms.*

Write the characteristic as

$$\begin{bmatrix} 3 & 4 & 5 & 2 \\ 2 & 1 & & \\ & 1 & & \end{bmatrix},$$

rows referring to constituents of invariant factors, columns to distinct latent roots, and elements to distinct elementary divisors.

There are four distinct roots $\alpha, \beta, \gamma, \delta$ say, in this order. The classical form has diagonal

$$C_3(\alpha),\ C_2(\alpha),\ C_4(\beta),\ C_1(\beta),\ C_1(\beta),\ C_5(\gamma),\ C_2(\delta).$$

The order is $n = 3 + 2 + 4 + 1 + 1 + 5 + 2 = 18$. The invariant factors are

$$(\lambda - \alpha)^3 (\lambda - \beta)^4 (\lambda - \gamma)^5 (\lambda - \delta)^2, \ (\lambda - \alpha)^2 (\lambda - \beta), \ (\lambda - \beta).$$

The elementary divisors are $(\lambda - \alpha)^3$, $(\lambda - \alpha)^2$, $(\lambda - \beta)^4$, $(\lambda - \beta)$, $(\lambda - \beta)$, $(\lambda - \gamma)^5$, $(\lambda - \delta)^2$. The rational form has diagonal B_{14}, B_3, B_1, say.

If all $\alpha, \beta, \gamma, \delta \neq 0$, the rank is 18. If $\alpha = 0$, the ranks of $C_3(\alpha)$, $C_2(\alpha)$, i.e.

$$\begin{bmatrix} \cdot & 1 & \cdot \\ \cdot & \cdot & 1 \\ \cdot & \cdot & \cdot \end{bmatrix}, \quad \begin{bmatrix} \cdot & 1 \\ \cdot & \cdot \end{bmatrix},$$

are 2, 1, and that of A is $2 + 1 + 4 + 1 + 1 + 5 + 2 = 16$. In general the rank is $n - \sigma$, where σ is the number of elementary divisors of type λ^i. (If $\beta = 0$, the rank is 15; if γ or $\delta = 0$, it is 17.)

6. A matrix equivalent to a diagonal matrix has a characteristic consisting solely of (possibly bracketed) units [1111 . . . 1]. It has linear elementary divisors.

7. To find a Real Classical Canonical Form for a Real Matrix.

Let the reduced characteristic function of A be resolved into *real* factors which are linear, when latent roots are real, and quadratic $(A^2 - pA - qI)$ when pairs of latent roots are conjugate complex numbers. Hence p and q are real, and $p^2 + 4q < 0$.

Corresponding to $(\lambda^2 - p\lambda - q)$ occurring to a power r in an invariant factor, we can take a modified simple classical submatrix

$$\begin{bmatrix} \cdot & 1 & \cdot & \cdot & \cdot & \cdot \\ q & p & 1 & \cdot & \cdot & \cdot \\ \cdot & \cdot & \cdot & 1 & \cdot & \cdot \\ \cdot & \cdot & q & p & 1 & \cdot \\ \cdot & \cdot & \cdot & \cdot & \cdot & 1 \\ \cdot & \cdot & \cdot & \cdot & q & p \end{bmatrix}$$

consisting of r pairs of rows (the illustration has three pairs) with $2r - 1$ units on the superdiagonal, and with q, p placed diagonally upon the even rows. The proof is analogous to that of Example 4. Write

$$\varphi(A) = (A^2 - pA - qI)^r \psi(A), \qquad v = u\psi(A),$$

where u has maximum grade. Now construct the chain $\xi_1 = vx$, $\xi_2 = vAx$, $\xi_3 = v(A^2 - pA - qI)x$, $\xi_4 = vA(A^2 - pA - qI)x$, $\xi_5 = v(A^2 - pA - qI)^2x$, $\xi_6 = vA(A^2 - pA - qI)^2x$. terminating with ξ_{2r}, since the next ξ contains the zero factor $\varphi(A)$. Let $\xi = Hx$, $y = Ax$, $\eta = Hy$, $\eta = HAH^{-1}\xi$. Then, proceeding as before, the corresponding section of the matrix HAH^{-1} is exhibited by

$$\begin{bmatrix} \eta_1 \\ \eta_2 \\ \eta_3 \\ \eta_4 \\ \vdots \end{bmatrix} = \begin{bmatrix} \cdot & 1 & \cdot & \cdot & \cdot & \cdot & \cdot & \cdots \\ q & p & 1 & \cdot & \cdot & \cdot & \cdot & \cdots \\ \cdot & \cdot & \cdot & 1 & \cdot & \cdot & \cdot & \cdots \\ \cdot & \cdot & q & p & 1 & \cdot & \cdot & \cdots \\ \cdot & \cdot & \cdot & \cdot & \cdot & \cdot & \cdot & \cdot \end{bmatrix} \begin{bmatrix} \xi_1 \\ \xi_2 \\ \xi_3 \\ \xi_4 \\ \vdots \end{bmatrix}.$$

8. The real classical form for the seven-rowed matrix, whose characteristic function is $(\lambda - 4)(\lambda^2 - 2\lambda + 3)^3$ and whose reduced characteristic function is $(\lambda - 4)(\lambda^2 - 2\lambda + 3)^2$, is

$$\begin{bmatrix} . & 1 & . & . & . & . & . \\ -3 & 2 & 1 & . & . & . & . \\ . & . & . & 1 & . & . & . \\ . & . & -3 & 2 & . & . & . \\ . & . & . & . & . & 1 & . \\ . & . & . & . & -3 & 2 & . \\ . & . & . & . & . & . & 4 \end{bmatrix}.$$

9. Matrices exist with prescribed invariant factors or elementary divisors, subject only to the field restrictions and the inequalities (22).

10. Show that the latent roots of the mth compound of A are the m-ary products $\prod_{i=1}^{m} a_{\nu_i}$ of the latent roots of A. (Rados, 1891.)

[If $C = HAH^{-1}$, $|H| \neq 0$, then $C^{(m)} = H^{(m)} A^{(m)} [H^{(m)}]^{-1}$, $|H^{(m)}| \neq 0$, by Binet-Cauchy and Sylvester. Examine the form of $C^{(m)}$.]

11. If A is non-singular, the latent roots of AB are the same as those of BA. More generally the latent roots of a product of matrices $ABC \dots K$, of which one matrix at most is singular, are the same as those of any other product of the same matrices in cyclic order.

$$[ABC \dots K = A(BC \dots KA)A^{-1}, \quad |A| \neq 0.]$$

12. If $f(x)$ is a scalar polynomial in x, the latent roots of the matrix $f(A)$ are $f(\lambda_i)$, where λ_i, $i = 1, 2, \dots, n$ are the latent roots of A. (Sylvester and Frobenius.)

[By Chapter IV, § 3, if $C = HAH^{-1}$, then $f(C) = H \,.\, f(A) \,.\, H^{-1}$. Thus (*ibid.* § 6) $f(C)$ and $f(A)$ have the same latent roots. Examine the form of $f(C)$.]

13. Extend the above to the function $f(A)/g(A)$, $|g(A)| \neq 0$.

6. Scalar Functions of a Square Matrix. Convergence.

For various reasons it is useful and interesting to extend the concept of a function of a matrix A to a wider class than polynomials or quotients of polynomials with scalar coefficients. If $f(z)$ is a power series in a complex variable z, convergent in a certain region of the complex plane of z, then the series can be used formally to *define* the corresponding matrix function $f(A)$, and we shall say that $f(A)$ converges provided that all of its n^2 elements converge. Each of these elements is itself a rather complicated infinite series, but the question of convergence is resolved by the aid of the identity $f(A) = H^{-1}f(C)H$. For (as on p. 7, Ex. 4) the diagonal of $f(C)$ contains elements $f(\lambda_i)$, and the superdiagonals of the submatrices contain derivatives up to $f^{(k-1)}(\lambda_i)$, where k is the highest exponent of the elementary divisors for the latent root λ_i, that is, k is the multiplicity of λ_i in the R.C.F.

of A. Also the elements of H are all finite, and so are those of H^{-1}. It follows, on multiplying out, that the condition of convergence of $f(A)$ is that the functions $f(\lambda_i), f'(\lambda_i), \ldots, f^{(k-1)}(\lambda_i)$ should converge for each λ_i. From the theory of infinite series we know that all of these functions diverge if λ_i lies outside the region of convergence of $f(z)$; that if λ_i lies within the region, the convergence of $f(\lambda_i)$ entails that of all derivatives $f^{(r)}(\lambda_i)$; and that if λ_i lies on the boundary of the region the earlier derivatives converge provided that the final one, here $f^{(k-1)}(\lambda_i)$, converges. These conditions, extended to each latent root, give the necessary and sufficient conditions for the convergence of a matrix power series.

In the same way scalar functions defined by *inverse factorial series* converging within "half-planes" have matrix analogues which converge, provided all of the latent roots lie within those half-planes; and matrix *continued fractions* can be defined, converging if none of the latent roots lies on any singular curve of the corresponding scalar continued fraction.

EXAMPLES

1. The matrix function e^A always converges.

2. The matrix $\log(1 + A)$, defined by the logarithmic series, converges if all the latent roots of A lie in or on the unit circle, excluding the point $z = -1$.

The theory of *convergence* of matrix functions with scalar coefficients can thus be founded on the classical canonical representation. A rather different approach towards the *expression* of the function is by way of the R.C.F. of A. Consider for example a matrix power series $f(A)$, the latent roots of A being within the region of convergence of $f(z)$, so that $f(A)$ converges. If the R.C.F. of A, let us say $\psi(A)$, is of degree p, then the matrix powers A^p, A^{p+1}, ..., are not merely congruent modulo $\psi(A)$ with certain polynomials of degree less than p, they are actually equal to those polynomials, since $\psi(A) = 0$. Consequently, by substituting these residual polynomials in the power series, we derive $g(A)$, a polynomial in A of degree less than p, which is in a certain sense the *residue* of the infinite matrix series $f(A)$ modulo $\psi(A)$, and the coefficients of which are convergent infinite series involving all the scalar coefficients in $f(z)$. *These scalar coefficients in* g (A), *though of complicated form, can be expressed by means of the recurrence relation corresponding to the reduced characteristic function.* The important fact is that a convergent function of a matrix A can be expressed as a *polynomial* in A, of degree less than p.

EXAMPLES

1. If $A^3 = c_0 I + c_1 A + c_2 A^2$, then $A^4 = c_0 c_2 I + (c_0 + c_1 c_2)A + (c_1 + c_2^2)A^2$. Find quadratic polynomials to express A^5, A^6.

2. If $\psi(A) = A^2 - A - I$, the residues mod ψ of $A^2, A^3, A^4, \ldots$, are $A + I$, $2A + I$, $3A + 2I, \ldots$, the coefficients of A and of I being the Fibonacci numbers 1, 1, 2, 3, 5, 8, ..., defined by the recurrence relation $y_{t+2} - y_{t+1} - y_t = 0$, $y_0 = 0$, $y_1 = 1$, or by the numerators and denominators of convergents to the continued fraction

$$1 + \frac{1}{1+}\;\frac{1}{1+}\;\frac{1}{1+}\;\cdots.$$

Prove that if $f(A) = (I - \frac{1}{3}A)^{-1}$, then $f(A) = \frac{3}{5}(A + 2I)$.

3. If a matrix A has latent roots α, β, γ with respective characteristics 3, 2, 1, and if α, β, γ lie within the region of convergence of a function $f(z)$, prove that a polynomial representation

$$f(A) = p_0 I + p_1 A + p_2 A^2 + \ldots + p_5 A^5$$

is possible, by comparing elements in the two forms of $f(C)$.

[(i) $f(C) = p_0 I + \ldots + p_5 C^5$: (ii) Compute $f(C)$ from III, p. 62. The comparison is then found to depend for consistency on a set of six linear equations in $p_0, p_1, \ldots, p_5$, the determinant of the equations being the confluent alternant $\Delta\{(\alpha\alpha\alpha)(\beta\beta)\gamma\}$, which is not zero.]

7. The Canonical Form of a Scalar Matrix Function.

The classical canonical form of a matrix function $f(A)$ can be deduced from that of A, provided that it is known which of the earlier derivatives $f'(z), f''(z), \ldots$ is the first to be non-zero for each latent root of A.

Theorem.—*If* $f'(\lambda) \neq 0$ *for each latent root* λ_i *of* A, *then the Segre characteristic of the matrix* f(A) *is the same as that of* A.

Proof.—If $C_p(a)$ is one of the simple classical submatrices, we have, for example when $p = 4$,

$$f[C_p(a)] = \begin{bmatrix} f(a) & f'(a) & \frac{1}{2!}f''(a) & \frac{1}{3!}f'''(a) \\ & f(a) & f'(a) & \frac{1}{2!}f''(a) \\ & & f(a) & f'(a) \\ & & & f(a) \end{bmatrix}. \qquad (23)$$

Now if $f'(a) \neq 0$, the later derivatives in the rows may be deleted by collineatory operations of Type II; then $f'(a)$ can be replaced by unity through operations of Type III, so that finally we have a simple classical submatrix similar to $C_p(\alpha)$. Since this can be done for each latent root the theorem is proved.

If for any latent root λ_i the first non-vanishing derivative is $f^{(k)}(\lambda_i)$, the modification in the characteristic for λ_i is readily obtained by the theory of chains. Suppose, for example, that $f'''(\lambda_i)$ is the first non-vanishing derivative, then the surviving non-diagonal elements in that submatrix after the first reduction are (1, 4), (2, 5), (3, 6), (4, 7), . . ., which fall into three chains indicated by 1, 4, 7, . . ., 2, 5, 8, . . ., 3, 6, 9 . . ., respectively. The general result is analogous and evident.

EXAMPLES

1. Discuss the characteristic of $f(A) = A^2 - 2A$, where the canonical form of A is

$$\begin{bmatrix} 1 & 1 & \cdot \\ & 1 & \cdot \\ & & 3 \end{bmatrix}.$$

For A, [21]; for $f(A)$, [(11)1]: $f'(1) = 0, f'(3) \neq 0$.

2. Do the same for the function $A^3 - 3A^2 + 3A$, where

$$C = \begin{bmatrix} 1 & 1 & \cdot & \cdot \\ & 1 & 1 & \cdot \\ & & 1 & \cdot \\ & & & 2 \end{bmatrix}.$$

8. Matrix Determinants: Sylvester's Interpolation Formula.

Consider the quotient of alternants $|\alpha^0 \beta^1 \gamma^2 \delta^3 f(\lambda)| / |\alpha^0 \beta^1 \gamma^2 \delta^3|$, as in Ex. 4, p. 63, where α, β, γ, δ are distinct and $f(\lambda)$ is a cubic polynomial. Since $f(\lambda)$ is a cubic the elements in the lowest row of the numerator determinant are a certain linear combination of those in earlier rows; thus the numerator is zero. The denominator, being the difference-product $\Delta(\alpha\beta\gamma\delta)$ of distinct variables, is not zero. By expanding the numerator in terms of the elements of its lowest row and observing that the co-factors of these are all difference-products, we obtain

$$f(\lambda) = \sum_{\alpha,\beta,\gamma,\delta} f(\alpha) \frac{(\lambda - \beta)(\lambda - \gamma)(\lambda - \delta)}{(\alpha - \beta)(\alpha - \gamma)(\alpha - \delta)}, \quad . \quad . \quad (24)$$

the well-known interpolation formula of Lagrange, by which the polynomial $f(\lambda)$ is determined from data $f(\alpha)$, $f(\beta)$, $f(\gamma)$, $f(\delta)$, and in general a polynomial of the nth degree is determined from $n + 1$ independent data.

Let us suppose next that certain variables coalesce, for example that β, γ, and δ coalesce to the one value β, the data now becoming $f(\alpha), f(\beta), f'(\beta), f''(\beta)$. We replace the former quotient of alternants by a quotient of *confluent* alternants, in this case by

$$| \alpha^0 \beta^1 2\beta^1 3\beta^1 f(\lambda) | / | \alpha^0 \beta^1 2\beta^1 3\beta^1 | .$$

The numerator vanishes for the same reason as before; the denominator does not, being the confluent difference-product $\Delta\{\alpha(\beta\beta\beta)\}$. If we proceed to expand the numerator in terms of its last row, we shall find that the co-factors of $f(\alpha)$ and $f(\beta)$ are confluent difference-products, but that those of the derivatives $f'(\beta)$ and $f''(\beta)$ are not quite so simple; none the less the expansion can be carried out, and gives a polynomial expression for $f(\lambda)$, with coefficients in which the data $f(\alpha), f(\beta), f'(\beta), f''(\beta)$ appear linearly. A similar procedure holds for any number of confluences of groups of variables.

We may apply the above results to the case of a polynomial $f(A)$ in a matrix A, relaxing one condition but adding another; namely, *the polynomial* f(A) *may be of any finite degree, but the variables* α, β, ... *in the alternant are to be the roots of the R.C.F. of* A.

It follows then, as in Ex. 4, p. 63, that for the case of different roots α, β, γ, δ we have

$$| \alpha^0 \beta^1 \gamma^2 \delta^3 f(A) | \equiv 0, \quad \text{mod } \Delta(\alpha\beta\gamma\delta A). \quad . \quad . \quad (25)$$

But
$$\Delta(\alpha\beta\gamma\delta A) = \Delta(\alpha\beta\gamma\delta)(A - \alpha I)(A - \beta I)(A - \gamma I)(A - \delta I)$$
$$= \Delta(\alpha\beta\gamma\delta)\,\phi(A) = 0. \quad . \quad . \quad . \quad . \quad (26)$$

Thus the matrix determinant in the numerator of our quotient vanishes as before, and expansion gives the result,

$$f(A) = \sum_{\alpha,\beta,\gamma,\delta} f(\alpha) \frac{(A - \beta I)(A - \gamma I)(A - \delta I)}{(\alpha - \beta)(\alpha - \gamma)(\alpha - \delta)}, \quad . \quad (27)$$

which is Sylvester's matrix analogue of Lagrange's formula. It is most important to observe that, in contrast with the scalar case, the polynomial $f(A)$ is of any finite degree, but that this increase of scope is balanced by the restriction that the scalar variables in the matrix formula must be the latent roots.

As might be supposed, when there are multiplicities among the latent roots of A the reduced $\psi(A) = 0$ takes the place of the characteristic equation $\phi(A) = 0$, and confluent alternants supplant the ordinary ones. We consider, for example,

$$| \alpha^0 \beta^1 2\beta^1 3\beta^1 f(A) | / | \alpha^0 \beta^1 2\beta^1 3\beta^1 | .$$

As in Ex. 6, p. 63, we have

$$|\,\alpha^0\beta^1 2\beta^1 3\beta^1 f(A)\,| = 0, \quad \text{mod } \Delta\{\alpha(\beta\beta\beta)A\}. \quad . \quad (28)$$

But
$$\Delta\{\alpha(\beta\beta\beta)A\} = \Delta\{\alpha(\beta\beta\beta)\}\,(A-\alpha)\,(A-\beta)^3$$
$$= (\beta-\alpha)^3\psi(A) = 0. \quad . \quad . \quad . \quad . \quad (29)$$

Thus the numerator of the quotient again vanishes, and expansion gives us the confluent analogue of Lagrange's formula, which we may conveniently leave in the form

$$(\beta-\alpha)^3 f(A) = -\begin{vmatrix} 1 & 1 & . & . & 1 \\ \alpha & \beta & 1 & . & A \\ \alpha^2 & \beta^2 & 2\beta & 1 & A^2 \\ \alpha^3 & \beta^3 & 3\beta^2 & 3\beta & A^3 \\ f(\alpha) & f(\beta) & f'(\beta) & \dfrac{f''(\beta)}{2!} & . \end{vmatrix}, \quad (30)$$

the general result being similar to this.

If this were all, not much would have been attained, for the reduced polynomial as given by such formulæ as the above is simply a disguised form of the residue $g(A)$ of $f(A)$, mod $\psi(A)$. But the formula holds also when $f(A)$ is a convergent matrix power series. In such a case we have $f(\lambda) = p_n(\lambda) + r_n(\lambda)$, where $p_n(\lambda)$ is the polynomial given by the first $n+1$ terms of $f(\lambda)$, and $r_n(\lambda)$ is the remainder after those terms. In the non-confluent case we have (on expanding the determinants by their final columns)

$$|\,\alpha^0\beta^1\gamma^2\delta^3 f(A)\,| = |\,\alpha^0\beta^1\gamma^2\delta^3 p_n(A)\,| + |\,\alpha^0\beta^1\gamma^2\delta^3 r_n(A)\,|$$
$$= |\,\alpha^0\beta^1\gamma^2\delta^3 r_n(A)\,|, \quad . \quad . \quad . \quad . \quad . \quad . \quad (31)$$

since, as we have seen, the first term vanishes. The surviving term also vanishes in the limit, since the last row in it has the elements $r_n(\alpha)$, $r_n(\beta)$, $r_n(\gamma)$, $r_n(\delta)$, $r_n(A)$; and the vanishing of these in the limit is simply the condition (p. 74) that $f(A)$ should converge. We conclude that Sylvester's formula is valid for convergent matrix power series: and similarly if A has p rows and columns.

In the confluent case we find likewise that the alternant in the numerator will vanish provided that in the limit $r_n(\alpha)$, $r_n(\beta)$, $r_n'(\beta)$, $r_n''(\beta)$, $r_n(A)$ tend to zero; but this again is exactly the condition (*ibid.*) that $f(A)$ should converge in the confluent case; and we conclude that the confluent interpolation formula is available.

EXAMPLES

1. If A is a non-singular matrix, the reciprocal matrix A^{-1} can be represented by a polynomial in A.

[Use the interpolation formula, or the R.C.F. A corresponding result holds for the adjoint matrix.]

2. The reciprocal of the characteristic matrix, $(A - \lambda I)^{-1}$, can be represented by a polynomial in A, provided that $|\lambda| > |\lambda_i|$ for each latent root λ_i.

3. Represent the function $(1 - \frac{1}{2}A)^{-1}$ by a polynomial, given that the R.C.F. $\psi(A) = A^2 - A - I$. [Cf. Ex. 2, p. 75.]

4. If $\psi(A) = A^2 - \alpha^2 I$, the polynomial for e^A is $\dfrac{\sinh\alpha}{\alpha} A + \cosh\alpha \, . \, I$.

5. *If* n *is a positive integer, the* n*th root of a matrix* A *can be expressed as a polynomial in* A.

9. The Segre Characteristic and the Rank of Matrix Powers.

As we have seen, the Segre characteristic of a matrix A, with respect to a certain latent root λ_i, consists of the first differences of the powers to which the scalar factor $(\lambda - \lambda_i)$ occurs in the characteristic determinant $|A - \lambda I|$ and in the H.C.F.'s of its minors of descending order. But the characteristic can also be determined from the first differences of the ranks of the powers of the matrix $A - \lambda_i I$. Consider, for example, a matrix A of the ninth order with two distinct latent roots α and β, the characteristic of α being (4, 2), that of β being (2, 1). The canonical form of A is thus

$$C = \begin{bmatrix} A_4 & & & \\ & A_2 & & \\ & & B_2 & \\ & & & B_1 \end{bmatrix}, \text{ where e.g. } \quad B_2 = \begin{bmatrix} \beta & 1 \\ . & \beta \end{bmatrix}. \qquad (32)$$

If we write $A_4 - \alpha I = U_4$, where I is the unit matrix of order 4, we observe that $(A_4 - \alpha I)^p$ has ranks 4, 3, 2, 1, 0, 0, . . . for $p = 0, 1, 2, 3, 4, \ldots$; similarly $(A_2 - \alpha I)^p$ has ranks 2, 1, 0, ... On the other hand the rank of $(B_q - \alpha I)^p$ is equal to q for all values of p, since the determinant has the value $(\beta - \alpha)^{pq} \neq 0$.

It will be more convenient to consider not the rank but its complement with respect to the order, that is, the *nullity*, $\nu = n - r$. The changes in the nullity in our example may then be represented by a diagram, thus:

```
| . . . .    . | . . .    . . | . .    . . . | .    . . . . |  (33)
| . .        . | .        . . |          . . |           . . |
 p=0, ν=0.      p=1, ν=2.      p=2, ν=4.      p=3, ν=5.   p=4, &c., ν=6.
```

The number of dots in a row represents the order of A_φ; and the increase of the α-nullity with increase of p is intuitively seen to be given by making vertical sections of the diagram, one place farther to the right each time. But the diagram is a Ferrers diagram of *partitions* of the integer 6, the total multiplicity of the latent root α. Hence we have the following theorem:

If the successive differences in nullity (or rank) in the matrix powers $(A - \lambda_i I)^p$, $p = 0, 1, 2, \ldots$, *are regarded as a partition (called the Weyr characteristic) of the total multiplicity of the latent root* λ_i, *then the conjugate partition is the Segre characteristic associated with* λ_i.

An alternative statement of the theorem may be given in terms of *dual* sets of numbers. Two sets of integers, for example [0, 1, 2, 6] and [2, 3, 4], are said to be *dual* (or *bicomplementary*) with respect to the complete set, here 0, 1, 2, . . ., 6, if the second set is made up of the *complements* with respect to $n + 1$ (here 7) of the integers *not included* in the first set. Thus in the example above the integers 3, 4, 5 do not appear in the first set; and their defects from 7, reversed, are 2, 3, 4. Since the first differences of the integers in dual sets are known to form conjugate partitions, we have the following theorem:

The nullities of $(A - \lambda_i I)^p$, $p = 0, 1, 2, \ldots$, *form a set of integers, the dual of which is the set formed by the powers to which* $(\lambda - \lambda_i)$ *occurs in the characteristic determinant of* A *and the H.C.F.'s of minors of descending order.*

EXAMPLES

1. Verify the dual nature of these sets for the illustrative example used in the demonstration.

2. The literal definition of conjugate partitions is this: If the integers a, b, c, d are in descending order of magnitude, and if $n = pa + qb + rc + sd$, where p, q, r, s are also positive integers, then the partition of n so derived is indicated by $(a^p b^q c^r d^s)$; the conjugate partition is $(\alpha^d\, \overline{\alpha - s}^{c-d}\, \overline{\alpha - s - r}^{b-c}\, \overline{\alpha - s - r - q}^{a-b})$, where $\alpha = p + q + r + s$. Show that this definition also leads to the theorem above.

10. **Historical Note.**—The form Γ was obtained by Jacobi in discussing bilinear forms uAx, which, in effect, he reduced to canonical form $v\Gamma y$; *J. für Math.*, **53** (1857), 265–70; or *Werke*, **3**, 585–90. The classical form C is first found* in C. Jordan, *Traité des Substitu-*

*The canonical form of a family of quadratics given by Weierstrass in 1868, discussed later in Chapter IX, is highly relevant, but the reductions there are congruent and the matrices symmetric.

tions (Paris, 1870), p. 114; the matrices there were of positive integers modulo p, the determinants involved being "non-singular", that is, prime to p. Frobenius and Landsberg, in papers cited in Chapter V, obtain the bilinear form vCy. Dickson has extended the result to any field; *Am. J. Math.*, **24** (1902), 101–8. The canonical reduction of bilinear forms has received new and thorough treatment recently by R. G. D. Richardson, *Trans. Am. Math. Soc.*, **26** (1924), 451–78.

The steps which we have used in reducing Γ to C were suggested in part by an elementary exposition of H. E. Hawkes, *Am. J. Math.*, **32** (1910), 101–14, but the prescription there given (p. 108) is inadequate without some further principle, such as the formation of chains.

The theorem usually credited to Sylvester, concerning the latent roots of a function of a matrix, was not given in full by him until 1883 (*Phil. Mag.* (5) **16**, p. 267), though he had given it for the square of a matrix in 1853. (See Muir's *History*, II, 123.) Frobenius gave it explicitly in 1877, *J. für Math.*, **84**, p. 11.

The formula of matrix interpolation was casually mentioned by Sylvester without proof; *Collected Works*, **4**, p. 111.

The deduction of the confluent from the simple case by differentiation was used by Jacobi (1825), *Werke*, **3**, 1–44, for the theory of partial fractions, which are obviously allied to the above interpolation formulæ, while the application to a matrix function $f(A)$ was given by Frobenius, *Berlin Sitzungsb.* (1896), 7–16.

The theorem on the convergence of a matrix power series is due to Ed. Weyr, *Bull. Sci. Math.*, **11** (1887), 205–15. The full statement for multiple latent roots is to be found in a paper by K. Hensel, *J. für Math.*, **155** (1926), 107–10.

Concerning alternants with differentiated columns we may refer to Muir's *History*, IV, p. 178, 201. It is a little remarkable that the alternant transformations of (5) and (8) above should arise from the inverted procedure of reducing the simpler form C to the more complicated form B by the vector-chain method of Chapter V.

The transformation of B into C by the alternant matrix was indicated by I. Schur, for the case of distinct latent roots, *Trans. Am. Math. Soc.*, **10** (1909), 159–175; and proved for the confluent case by A. C. Aitken, *Proc. Roy. Soc. Edin.*, **51** (1930), 81–90.

CHAPTER VII

Congruent and Conjunctive Transformations: Quadratic and Hermitian Forms

It was pointed out in Chapter IV that the linear transformation of the variables of a quadratic form $x'Ax$, or of an Hermitian form $\bar{x}'Ax$, involved in the one case the congruent transformation $H'AH$, in the other the conjunctive transformation $\overline{H}'AH$, the matrix A being respectively symmetric or Hermitian. It will be convenient to consider these special cases first, and to reserve till later the discussion of the congruent or conjunctive transformation of a general matrix A in the field.

1. The Congruent Reduction of a Conic.

The initial steps of the reduction of a quadratic form to a sum of terms involving squares only (as commonly phrased, "to a sum of squares"), which is the same process as the congruent reduction of the symmetric matrix A of the form to diagonal canonical shape, depend upon whether the matrix A has non-zero elements in the diagonal or not. The leading features of the reduction may be illustrated by a well-known example, that of referring a conic

$$\phi = ax^2 + by^2 + cz^2 + 2fyz + 2gzx + 2hxy = 0 \qquad (1)$$

to a self-conjugate triangle of reference, the co-ordinates being homogeneous. For first vertex of the conjugate triangle any point R not lying on the conic may be taken. For the second, any point S lying on the polar of R but not lying on the conic may be taken; and for third, T, the pole of RS. Evidently RST is a self-conjugate triangle; for it is readily verified that each vertex is the pole of the opposite side. In the general case there are ∞^2 ways of choosing R, ∞^1 ways of choosing S, and ∞^3 ways of choosing the triangle. [The number of degrees of freedom of choice for n variables would be $\frac{1}{2}n(n-1)$.]

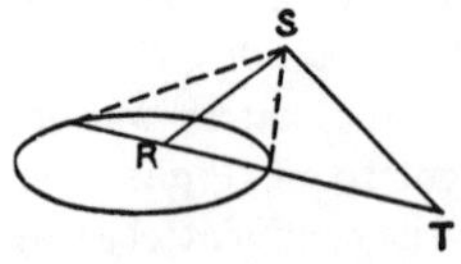

Analytically, if the coefficient $a \neq 0$, then the vertex $\{1, 0, 0\}$ of the triangle of reference does not lie on the conic. We take this vertex as the point R, its polar being $ax + hy + gz = 0$. In ϕ we collect all terms involving x into one perfect square expression, and write (following the conventional notation)

$$\phi = \{(ax + hy + gz)^2 + Cy^2 - 2Fyz + Bz^2\}/a, \quad . \quad (2)$$

where $C = ab - h^2$, &c.

Again, if $\Delta = \begin{vmatrix} a & h & g \\ h & b & f \\ g & f & c \end{vmatrix}$, and $C \neq 0$, we proceed similarly with the remaining terms, writing

$$Cy^2 - 2Fyz + Bz^2 = \{(Cy - Fz)^2 + a\Delta z^2\}/C, \quad . \quad (3)$$

since $BC - F^2 = a\,\Delta$. Lastly, if $\Delta \neq 0$, this brings ϕ to the form

$$\phi = \alpha_1 \xi_1{}^2 + \alpha_2 \xi_2{}^2 + \alpha_3 \xi_3{}^2,$$

$$\text{where} \quad \left.\begin{array}{lll} \xi_1 = ax + hy + gz, & \xi_2 = Cy - Fz, & \xi_3 = \Delta z, \\ \alpha_1 = 1/a, & \alpha_2 = 1/aC, & \alpha_3 = 1/C\Delta. \end{array}\right\} \quad . \quad . \quad (4)$$

Such a process, which is rational and terminating, can obviously be extended to quadratic forms in n variables, provided that at each stage the square of some unreduced variable has a non-zero coefficient: and the example illustrates the first and leading feature of reduction to congruent canonical form.

A second feature arises when the conic circumscribes the triangle of reference, so that $a = 0$, $b = 0$, $c = 0$. In this case the above argument breaks down; but since the coefficient of at least one product term, let us say h, is non-zero, we may use the triangle $z = 0, x + y = 0$, $x - y = 0$ as a new triangle of reference, one of whose vertices is not on the conic, and the original method may be continued. The reduction applies however specialized the conic may be.

2. The Symmetrical Bilinear Form.

If we subject a quadratic form $\phi = x'Ax$ to the polarizing operation $\frac{1}{2}\sum_{i=1}^{n} y_i \frac{\partial}{\partial x_i}$ we obtain the symmetrical bilinear form

$$y'Ax = \sum_{i,j=1}^{n} y_i a_{ij} x_j, \qquad a_{ji} = a_{ij}. \quad . \quad . \quad . \quad (5)$$

which can also be written $x'Ay$. If we regard x and y as points, then $\phi = 0$ is the point equation of a quadric in $n-1$ dimensions, and $y'Ax = 0$ is that of the *polar* of y with respect to the quadric. If ξ, η denote the points x, y referred to a new non-degenerate simplex we have $x = H\xi$, $y = H\eta$, $|H| \neq 0$, and the transformed quadric is $\xi' B\xi = 0$, where $B = H'AH$, as we have seen: consequently, for the polar bilinear form,

$$y'Ax = \eta' H'AH\xi = \eta' B\xi = \tfrac{1}{2}\sum_i \eta_i \frac{\partial}{\partial \xi_i}\{\sum_{i,j} \xi_i b_{ij} \xi_j\}. \quad . \quad (6)$$

We see that whether we polarize first and change the variables next, or *vice versa*, the result is the same. Polarization is therefore an invariant process in the group; and the theory of symmetrical bilinear forms is no more general than that of the corresponding quadratic forms.

3. Generalized Quadratic Forms and Congruent Transformations.

By a simple device of notation, L. E. Dickson has shown * that most of the properties of quadratic forms and Hermitian forms, and indeed of more general forms including both, can be considered simultaneously. Following him we shall write

$$\left.\begin{aligned} A' &= \tilde{A} = [\tilde{a}_{ij}], \qquad B' = \tilde{B} = [\tilde{b}_{ij}], \\ B &= \tilde{H}'AH, \qquad |H| \neq 0, \end{aligned}\right\} \quad . \quad . \quad . \quad (7)$$

and interpret the sign $\sim$ as follows: either the matrices A and B are Hermitian, and the $\sim$ has the significance of the bar-sign denoting conjugacy, or else the sign $\sim$ may be ignored and the matrices are symmetrical. In this way two distinct but parallel sets of theorems are covered by one notation, which extends to the forms themselves, so that the Hermitian or quadratic form is now written

$$\tilde{x}'Ax = \sum_{i,j=1}^{n} \tilde{x}_i a_{ij} x_j, \qquad a_{ij} = \tilde{a}_{ji}, \qquad . \quad . \quad . \quad (8)$$

and the linear transformations are $x = H\xi$, $\tilde{x} = \tilde{H}\tilde{\xi}$. We shall call such a form a *generalized* quadratic form. The device takes account of the case when the quadratic form has *complex* coefficients, and is not merely a particular case of an Hermitian form with real coefficients. Indeed it is a remarkable fact—elementary though it be—that an Hermitian form is always *real*, in spite of its complex matrix.

* *Modern Algebraic Theories* (Chicago, 1926), Chapter IV, p. 68.

This follows from the identity $\bar{x}'Ax = x'A'\bar{x}$, or from examination of conjugate terms in the form.

The elementary operations of congruent transformation given in Chapter II are readily adapted to the present convention: for example, the congruent operation of Type II is given by

$$H = I + (h)_{ij},$$

$$\bar{H}'AH: \ \text{row}_j + \bar{h}\,\text{row}_i, \qquad \text{col}_j + h\,\text{col}_i. \quad . \quad . \quad (9)$$

For instance

$$\begin{bmatrix} 1 & \bar{q} \\ . & 1 \end{bmatrix} \begin{bmatrix} a_{11} & a_{12} \\ a_{21} & a_{22} \end{bmatrix} \begin{bmatrix} 1 & . \\ q & 1 \end{bmatrix} = \begin{bmatrix} a_{11}+\bar{q}a_{21}+qa_{12}+\bar{q}qa_{22} & a_{12}+\bar{q}a_{22} \\ a_{21}+qa_{22}, & a_{22} \end{bmatrix}.$$

The operation of Type I—the simultaneous interchange of the same rows as columns—is also available, being, as was observed, congruent as well as collineatory.

4. The Rational Reduction of Quadratic and Hermitian Forms.

Theorem I.—*Any generalized quadratic form* $\bar{x}'Ax$ *can be reduced by a congruent (or conjunctive) transformation in a field* $\mathcal{F}$ *to the sum of* r *non-zero terms*

$$\alpha_1\bar{\xi}_1\xi_1 + \alpha_2\bar{\xi}_2\xi_2 + \dots + \alpha_r\bar{\xi}_r\xi_r, \qquad \bar{\alpha}_i = \alpha_i \neq 0.$$

Proof.—(i) If $a_{11} = 0$ but $a_{ii} \neq 0$, then a_{ii}, which is equal to $\bar{a}_{ii}$, can be brought to the leading position by an interchange of Type I. (ii) If all $a_{ii} = 0$ and $a_{ij} \neq 0$ $(i < j)$, the operation $\text{row}_i + a_{ij}\,\text{row}_j$, $\text{col}_i + \bar{a}_{ij}\,\text{col}_j$, places a non-zero element, $2\bar{a}_{ij}a_{ij}$, at the position a_{ii}, whence it can be brought as above to a_{11}. (iii) If all a_{ij} vanish, then $A = 0$, $r = 0$, and the theorem is already true but trivial.

(iv) Now consider the matrix product

$$\bar{H}_1'A\,H_1 = \begin{bmatrix} 1 & . \\ -\bar{b}'a_{11}^{-1} & I \end{bmatrix} \begin{bmatrix} a_{11} & b \\ \bar{b}' & A_1 \end{bmatrix} \begin{bmatrix} 1 & -a_{11}^{-1}b \\ . & I \end{bmatrix}, \quad (10)$$

where the generalized symmetric matrix A has been partitioned, A_1 being a submatrix of the same type of order $n - 1$, and b denoting the row vector $[a_{12}, a_{13}, \dots, a_{1n}]$. The product is seen to be

$$\begin{bmatrix} a_{11} & . \\ . & A_1 - \bar{b}'a_{11}^{-1}b \end{bmatrix}. \quad . \quad . \quad . \quad . \quad . \quad (11)$$

Thus a_{11} has been isolated by a conjunctive transformation, and $A_1 - \tilde{b}'a_{11}^{-1}b$ is again generalized symmetric.

(v) The remainder of the proof is obvious. As in previous cases, r successive real elements may be isolated in this way down the diagonal until either all remaining $a_{ij} = 0$ $(i > r,\ j > r)$ or else until $r = n$. By combining all the operations, we have found, in a terminating sequence of rational steps, the identity

$$B = \tilde{H}'AH = \begin{bmatrix} a_1 & & & & & \\ & a_2 & & & & \\ & & \cdot & & & \\ & & & \cdot & & \\ & & & & a_r & \\ & & & & & \cdot \\ & & & & & & \cdot \end{bmatrix}, \quad |H| \neq 0, \qquad \tilde{a}_i = a_i. \quad (12)$$

Inserting the variables we obtain the identity

$$\tilde{x}'Ax = \tilde{\xi}'B\xi = a_1\tilde{\xi}_1\xi_1 + a_2\tilde{\xi}_2\xi_2 + \ldots + a_r\tilde{\xi}_r\xi_r, \quad . \quad (13)$$

where $x = H\xi$, $\tilde{x} = \tilde{H}\tilde{\xi}$. Thus the theorem is proved.

5. The Rank of a Quadratic or Hermitian Form.

The number r of terms in the above canonical form is called the *rank* of the form. By the theorem of invariance of rank under equivalent transformations (Chapter III, p. 27) we deduce at once that in a given field any generalized quadratic form of rank r may be reduced to the form $\sum_{i=1}^{r} a_i\tilde{\xi}_i\xi_i$, where r is invariant, and each coefficient a_i is real and non-zero. So far the reductions have been rational, but if we merge the coefficient in the variable by a further real linear transformation $z_i = |a_i|^{\frac{1}{2}}\xi_i$, involving $|a_i|$, the modulus of a_i, we may reduce any generalized quadratic form of rank r to a sum of r terms $\tilde{z}_i z_i$ with coefficients ± 1. If p of the terms are positive and therefore $r - p$ negative, we may permute the variables if necessary, and finally obtain the reduced form

$$\tilde{x}'\tilde{P}'APx = \tilde{z}_1z_1 + \tilde{z}_2z_2 + \ldots + \tilde{z}_pz_p - \tilde{z}_{p+1}z_{p+1} - \ldots - \tilde{z}_rz_r. \quad (14)$$

EXAMPLES

1. A real quadratic form of rank r may be reduced by real transformations to $z_1^2 + z_2^2 + \ldots + z_p^2 - z_{p+1}^2 - \ldots - z_r^2$.

2. An Hermitian form may be reduced by conjunctive transformations to $\bar{z}_1z_1 + \bar{z}_2z_2 + \ldots + \bar{z}_pz_p - \bar{z}_{p+1}z_{p+1} - \ldots - \bar{z}_rz_r$.

3. A symmetric bilinear form may be reduced to

$$y_1z_1 + y_2z_2 + \dots + y_pz_p - y_{p+1}z_{p+1} - \dots - y_rz_r.$$

6. The Congruent Reduction of a Skew Bilinear Form.

With very little change the method of § 4 may be used to effect the rational reduction by congruent or conjunctive transformations of a skew symmetric, or skew Hermitian, matrix $S = [s_{ij}] = -\widetilde{S'}$.

Theorem II.—*A skew bilinear form* u S x, *where* S′ = —S, *must be of even rank* r = 2k, *and can be reduced by a congruent transformation, rational in the field of* S, *to the form*

$$uSx = vH'SHy = v_1y_2 - v_2y_1 + v_3y_4 - v_4y_3 + \dots + v_{r-1}y_r - v_ry_{r-1}.$$

Proof.—Let the generalized skew matrix $S = -\tilde{S}'$ be partitioned into submatrices S_{ij} of order 2×2, where, for example,

$$S_{11} = \begin{bmatrix} s_{11} & s_{12} \\ -\tilde{s}_{12} & s_{22} \end{bmatrix} = -\tilde{S}_{11}', \quad S_{12} = \begin{bmatrix} s_{13} & s_{14} \\ s_{23} & s_{24} \end{bmatrix} = -\tilde{S}_{21}'. \quad (15)$$

If S is of odd order we may provisionally add a last row and column of zeros for this purpose. Then the cases (i), (ii), and (iii) of Theorem I can be dealt with for the submatrices S_{ij} just as they were for the elements a_{ij}. We can ensure therefore that the leading submatrix S_{11} may be made non-zero, and we form the product

$$\tilde{H}_1'SH_1 = \begin{bmatrix} I & . \\ \tilde{T}'S_{11}^{-1} & I \end{bmatrix} \begin{bmatrix} S_{11} & T \\ -\tilde{T}' & S_1 \end{bmatrix} \begin{bmatrix} I & -S_{11}^{-1}T \\ . & I \end{bmatrix}, \quad (16)$$

where S has been partitioned, the submatrix S_1 being skew of order $n - 2$, and T denoting a submatrix of order $2 \times (n - 2)$. The transformation is conjunctive, for

$$-(\widetilde{S_{11}^{-1}T})' = -\tilde{T}'(\tilde{S}_{11}')^{-1} = \tilde{T}'S_{11}^{-1}, \text{ since } (\tilde{S}_{11}')^{-1} = -S_{11}^{-1}. \quad (17)$$

The product is found to be

$$\tilde{H}_1'SH_1 = \begin{bmatrix} S_{11} & . \\ . & S_1 + \tilde{T}'S_{11}^{-1}T \end{bmatrix}. \quad . \quad . \quad (18)$$

Thus S_{11} has been isolated, and the submatrix $S_1 + \tilde{T}'S_{11}^{-1}\,T$ is again generalized skew symmetric, a fact which also follows from the conjunctive nature of the transformation. From this point we proceed exactly as in Theorem I, until we exhaust the non-zero submatrices in the diagonal. By combining the various operations we have the identity

$$\Sigma = \tilde{K}'SK = \begin{bmatrix} \Sigma_1 & & & & \\ & \Sigma_2 & & & \\ & & \cdot & & \\ & & & \cdot & \\ & & & & \Sigma_k \end{bmatrix}; \quad |K| \neq 0, \quad . \quad . \quad (19)$$

where, in the skew symmetric case, we can write $\Sigma_i = \begin{bmatrix} \cdot & \sigma_i \\ -\sigma_i & \cdot \end{bmatrix}$, and $\sigma_i \neq 0$. By final operations in the field, σ_i^{-1} row_{2i-1}, σ_i^{-1} col_{2i-1}, we obtain the skew symmetric form,

$$H'S\,H = \begin{bmatrix} U-U' & & & \\ & U-U' & & \\ & & & \\ & & & U-U' \\ & & & & \cdot \end{bmatrix}, \quad . \quad . \quad (20)$$

where there are k submatrices $U - U'$, and $U = \begin{bmatrix} \cdot & 1 \\ \cdot & \cdot \end{bmatrix}$. Inserting the variables v and y, where $u = v\,H'$, $x = H\,y$, we obtain the reduced bilinear form as desired,

$$uSx = vH'SHy = v_1y_2 - v_2y_1 + \ldots + v_{r-1}y_r - v_ry_{r-1}. \quad (21)$$

The proof shows that the rank of a skew symmetric bilinear form is necessarily even. This is also an immediate consequence of the well-known theorem of determinants that a skew symmetric determinant of odd order vanishes identically.

The conjunctive reduction of a complex skew Hermitian matrix, as distinct from a skew symmetric one, follows the same lines as the above, diverging only at stage (20), where submatrices of type

$$\begin{bmatrix} i\alpha & 1 \\ -1 & i\beta \end{bmatrix}$$

are obtained, α and β being real. The nature of the rational canonical bilinear form is easily inferred. If irrationalities are permitted, we may further reduce the submatrix conjunctively by

$$|\alpha|^{-\frac{1}{2}}\,\text{row}_1, \quad |\alpha|^{-\frac{1}{2}}\,\text{col}_1, \quad |\beta|^{-\frac{1}{2}}\,\text{row}_2, \quad |\beta|^{-\frac{1}{2}}\,\text{col}_2,$$
$$\text{row}_2 \pm i\,|\alpha\beta|^{-\frac{1}{2}}\,\text{row}_1, \quad \text{col}_2 \mp i\,|\alpha\beta|^{-\frac{1}{2}}\,\text{col}_1,$$

(cf. (9), p. 85), if $\alpha \neq 0$, $\beta \neq 0$, thus obtaining a canonical form which is purely diagonal but also purely imaginary. This final result follows rationally if one of α, β is zero.

7. Definite and Indefinite Forms. Sylvester's Law of Inertia.

It was first discovered by Sylvester, and independently a few years later by Jacobi, that in real linear transformations of a real quadratic form not only the rank r but *the index of positiveness* p remain *invariant*—a theorem picturesquely named by Sylvester the *Law of Inertia* of quadratic forms.

Theorem III.—*In whatever way a real quadratic form is reduced by real linear transformations to a sum of positive and negative squares, the numbers* p *of positive terms and* r — p *of negative terms remain the same, provided that the squares are real and independent.*

Proof.—Suppose that the form $x'Ax$ of rank r has been reduced as above to the two different expressions

$$\left.\begin{array}{ll} & y_1^2 + y_2^2 + \ldots + y_p^2 - y_{p+1}^2 - \ldots - y_r^2 \\ \text{and} & z_1^2 + z_2^2 + \ldots + z_q^2 - z_{q+1}^2 - \ldots - z_r^2. \end{array}\right] \quad . \quad . \quad (22)$$

We shall prove that $p = q$. If possible let $p > q$. Since the variables y_i and z_i are certain homogeneous linear functions of the x_i with coefficients in the real field, it follows that the expressions (22) are equal for all values of the x_i. Now the $n - p + q$ homogeneous equations

$$z_1 = 0, \quad z_2 = 0, \ \ldots, \ z_q = 0, \quad y_{p+1} = 0, \ \ldots, \ y_n = 0, \quad (23)$$

being fewer than n in number, can be satisfied (Chapter III, p. 29) by solutions $x_i = X_i$, $i = 1, 2, \ldots, n$ which are not all zero. If Y_i and Z_i denote the corresponding values of y_i and z_i, we obtain by equating the expressions (22)

$$Y_1^2 + Y_2^2 + \ldots + Y_p^2 + Z_{q+1}^2 + \ldots + Z_r^2 = 0.$$

But each term on the left is necessarily positive or zero: hence each must be zero, and in particular $Y_1, Y_2, \ldots, Y_p$ are zero. We have therefore found a set of solutions X_i, not all zero, for the n equations

$$y_i = 0, \qquad i = 1, 2, \ldots, n.$$

But this contradicts the condition that the y_i are linearly independent functions of the x_i. Hence q cannot be less than p. Similarly p cannot be less than q: so that $p = q$ and the theorem is proved.

A similar proof holds for Hermitian or for generalized forms, z_i^2, &c., being replaced by $\tilde{z}_i z_i$.

The integer p is called the *index* of the real quadratic or of the Hermitian form; it is thus an *integer invariant* of the forms. Sometimes a different index, namely the difference between the number of positive and the number of negative terms, $p - (r - p)$, or $2p - r$, is used instead; this invariant is called the *signature*.

When the canonical form of the quadratic has no negative terms, and is of full rank n, so that $p = n$, the form, whether quadratic or Hermitian, is said to be *positive definite*. When $p = r$, $r < n$, it is often called *positive semi-definite*, but more strictly may be called *non-negative definite of rank* r. When $p = r = 0$, the form vanishes identically and may be called *zero definite* or *null*. When $p = 0$, $r > 0$, the canonical form has negative terms only, and is called *negative definite*, or *non-positive definite of rank* r, as the case may be. The distinction between positive definite and non-negative definite of rank $r < n$ is not a pedantic one; it has an important bearing on questions of maxima and minima in functions of many variables, stability in small oscillations, redundant variables in statistical problems, and the like.

8. **Determinantal Theorems concerning Rank and Index.**

In many applications of the theory of quadratic forms it is useful to possess criteria by which we can recognize whether a quadratic (or Hermitian) form is definite or not. Several such criteria can be derived by examining the sign of the principal minors of the symmetric (or Hermitian) determinant $|A|$.

Some of the theorems exemplify the properties of compound matrices mentioned in Chapter III, p. 28. By taking the mth compound of the identity (12) above, we obtain without difficulty

$$(\widetilde{H^{(m)}})' A^{(m)} H^{(m)} = B^{(m)}, \quad |H^{(m)}| \neq 0, \quad . \quad . \quad (24)$$

where $B^{(m)}$, easily computed from B itself, is a diagonal matrix, with leading element $a_1 a_2 \ldots a_m$, and is therefore in canonical form; which shows that the canonical transformation of a generalized quadratic form is reflected by a corresponding transformation of a class of *compound quadratics*, for $m = 2, 3, \ldots, n - 1$. The last of these is well known as the *adjoint form*; its matrix, apart from a permutation of rows and columns, is the adjoint of A.

For example, if $B = \begin{bmatrix} \alpha_1 & & & \\ & \alpha_2 & & \\ & & \alpha_3 & \\ & & & . \end{bmatrix}$, then $B^{(2)} = \begin{bmatrix} \alpha_1\alpha_2 & & & & & \\ & \alpha_1\alpha_3 & & & & \\ & & \alpha_2\alpha_3 & & & \\ & & & . & & \\ & & & & . & \\ & & & & & . \end{bmatrix}$ and

$B^{(3)} = \begin{bmatrix} \alpha_1\alpha_2\alpha_3 & & & \\ & . & & \\ & & . & \\ & & & . \end{bmatrix}$, with respective orders 4, 6, 4 and ranks 3, 3, 1.

Theorem IV.—*The principal minors of the determinant* $|\,A\,|$ *of a generalized quadratic form which is positive definite are all positive.*

Proof.—By the initial step in Theorem I, the first term in the reduction of $x'Ax$ was obtained as $a_{11}\tilde{\xi}_1\xi_1$. Since the form $\tilde{x}'Ax$ is positive definite the element a_{11} must be positive. Similarly, by congruent interchange, every a_{ii} must be positive.

Next, since the diagonal elements of the canonical form B are positive real numbers α_i, those of the mth compound $B^{(m)}$, being the m-ary products of the α_i, are also positive and real. By (24) above it follows that the mth compound quadratic form, of matrix $A^{(m)}$, is positive definite, of full rank $\binom{n}{m}$. Hence, as we have just shown, the diagonal elements of $A^{(m)}$ are all positive. But these are simply the principal minors of order m of the determinant $|\,A\,|$. Hence the theorem is proved.

For negative definite forms we see similarly that the mth compounds $B^{(m)}$ have entirely positive or entirely negative elements in the diagonal according as m is even or odd, so that in this case the principal minors of $|\,A\,|$ are positive if of even order, negative if of odd order. For non-negative definite forms it can be seen that principal minors may take the value zero as well as positive values.

A necessary and sufficient condition for a Hermitian or a real symmetric matrix A to be positive definite is that the n leading principal minors of A, of orders $1, 2, 3, \ldots, n$, should all be positive. For this implies that $\alpha_1, \alpha_1\alpha_2, \ldots$,—the n corresponding leading minors of the compound matrices $B^{(m)}$—are all positive. In the negative definite case the signs of the leading minors are alternately $-, +$: and in the general case, changes of sign in the sequence furnish a means of determining the signature.

Various criteria may also be given for simplifying the work of finding the rank of a matrix A.

Theorem V.—*If in a general matrix* A (*not necessarily symmetric or otherwise specialized; not even square*) *a certain minor of order* r *is non-zero while all minors of order* r + 1 *which contain this particular minor are zero, then the matrix* A *is of rank* r.

Proof.—The proof rests on the fundamental identity of Sylvester, as given in *Invariants*, p. 45, II. For example, adopting the notation for determinants there used, let $[\alpha\beta\gamma\delta\epsilon\zeta]_{123}$ be a matrix of order 3×6, such that the determinant $(\alpha\beta)_{12} \neq 0$, while all $(\alpha\beta\xi) = 0$, $\xi = \gamma, \delta, \epsilon, \zeta$. Let ω denote the *column* $\{0, 0, 1\}$; then

$$(\alpha\beta\omega)(\gamma\delta\epsilon) = (\dot{\gamma}\dot{\delta}\omega)(\alpha\beta\dot{\epsilon}), \quad . \quad . \quad . \quad . \quad (25)$$

identically, the dots indicating (*loc. cit.*) that the indices affected receive determinantal permutation and sign. The three terms so indicated on the right vanish since $(\alpha\beta\xi) = 0$.

Hence the minor $(\gamma\delta\epsilon) = 0$, since $(\alpha\beta)_{12} = (\alpha\beta\omega) \neq 0$. . (26)

Hence in general all minors of order $r + 1$ which have r *rows in common with the given minor* vanish, as also must those which have r columns in common with it.

Next, consider the Sylvester identity involving products of six-rowed determinants

$$(\alpha\beta 3456)(\gamma\delta\epsilon 126) = (\dot{\gamma}\dot{\delta}3456)(\alpha\beta\dot{\epsilon}\dot{1}\dot{2}6), \quad . \quad . \quad (27)$$

where the $(3) \equiv \{0, 0, 1, 0, 0, 0\}$, $(4) \equiv \{0, 0, 0, 1, 0, 0\}$, and so on, and $n = 6$. Since by (25) all minors containing $(\alpha\beta)_{ij}$ vanish, the right-hand side of (27) vanishes. Hence from the left-hand side, $(\gamma\delta\epsilon)_{345} = 0$.

The same treatment applies to a minor with rows or columns in common with those of the original non-zero minor. For instance, γ, δ, ϵ may include α or β.

The proof, summarized by this example, is general. Thus all minors of orders $r + 1$ are zero; and since one of order r is non-zero the rank of A is r.

The next theorem is more special, referring to a symmetric or Hermitian matrix A.

Theorem VI.—*If in a symmetric* (*or Hermitian*) *matrix* A *a certain* principal minor *of order* r *is non-zero, and all* principal minors *of orders* r + 1 *and* r + 2 *containing it are zero, then the matrix* A *is of rank* r.

Proof.—The minor in question may be moved to leading position. Let it be $(\alpha\beta\gamma)_{123}$, and let $(\alpha\beta\gamma\delta)_{1234}$ and $(\alpha\beta\gamma\delta\epsilon)_{12345}$ be two vanishing minors. Then we have the determinantal identity

$$(\alpha\beta\gamma\delta\epsilon)_{12345}\,(\alpha\beta\gamma)_{123} = \begin{vmatrix} (\alpha\beta\gamma\delta)_{1234} & (\alpha\beta\gamma\epsilon)_{1234} \\ (\alpha\beta\gamma\delta)_{1235} & (\alpha\beta\gamma\epsilon)_{1235} \end{vmatrix}. \tag{28}$$

Since $(\alpha\beta\gamma\delta)_{1234} = 0$, we have $(\alpha\beta\gamma\epsilon)_{1234}\cdot(\alpha\beta\gamma\delta)_{1235} = 0$. But by symmetry, or the Hermitian property, the two factors are either identical or conjugate. Hence we have $(\alpha\beta\gamma\delta)_{1235} = 0$, so that any minor of order $r + 1$, not necessarily principal, containing $(\alpha\beta\gamma)_{123}$ is zero. Hence, by Theorem V, the rank of A is r.

Note.—If all diagonal elements a_{ii} are zero, and all diagonal minors of the second order $|\, a_{ii}\, a_{jj} \,|$ are zero, the matrix A is the null matrix; for then $\bar{a}_{ij}a_{ij} = 0$, $i \neq j$, and so $a_{ij} = 0$.

Corollary.—*By a permutation* Ω *of the rows and columns of a symmetric or Hermitian matrix* A *of rank* r, *it is possible to ensure that the leading minor of order* r *is non-zero, and that no consecutive two of the leading minors of orders* 1, 2, . . . , r — 1 *vanish together.*

If all $a_{ii} = 0$ and all $|\, a_{ii}\, a_{jj} \,| = 0$, then, as we have seen, the matrix A is null. If, however, the rank r is non-zero, let us suppose that the longest sequence of leading minors obeying the above conditions breaks down in such a way that $A_p = |\, a_{11}a_{22} \ldots a_{pp} \,| \neq 0$, $A_{p+1} = 0$, $A_{p+2} = 0$, $p < r$. This would mean that all principal minors of orders $p + 1$ and $p + 2$ containing A_p were necessarily zero, so that, by Theorem V, the rank would be p, contrary to hypothesis. Hence p cannot be less than r; and of course it cannot exceed r. Hence $p = r$, which proves the Corollary.

Theorem VII.—*A symmetric* (*or Hermitian*) *matrix of non-zero rank* r *contains at least one non-zero* principal *minor of order* r.

Proof.—Since the rank is r, there is at least one set of r linearly independent rows. By symmetry, or symmetry and conjugacy in the Hermitian case, the corresponding set of r columns is linearly independent. Hence the minor common to these rows and columns, a principal minor, must be non-zero.

This result is of course part of the Corollary of Theorem VI, but is important enough to receive simple independent proof.

The fact is also obvious from a consideration of compound matrices

The rth compound of A, namely $A^{(r)}$, must be of rank 1, for its canonical form $B^{(r)}$ has a single leading non-zero element $a_1 a_2 \dots a_r$, and all other elements 0. Now $A^{(r)}$ must have at least one non-zero element in the diagonal; for if all its diagonal elements were zero there would be at least one non-vanishing principal minor of order 2. This cannot be, since the rank is 1. Hence $A^{(r)}$ has at least one non-zero diagonal element, and this is a principal minor of order r in A.

It is worth note that, if A is Hermitian of rank r, the rank of $A^{(m)}$ is

$$\binom{r}{m} = r(r-1)\ (r-2) \dots (r-m+1)/m! \qquad (29)$$

9. Congruent Reduction of a General Matrix to Canonical Form.

If in (10) above we discard the $\sim$ sign and allow the matrix A to be perfectly general in the field, we obtain an identity involving $H_1'AH_1$ which does not wholly isolate the leading element a_{11}, but *semi-isolates* it. We can, however, proceed in the same way as before, with semi-isolation of diagonal elements at each stage, and thereby, in at most $n-1$ such steps, derive by congruent transformation a Jacobian form (transposed from the previous one, p. 64)

$$H'AH = [a_{ij}], \qquad (a_{ij} = 0, \quad i < j). \quad . \quad . \quad . \quad (30)$$

If r is the rank of A we shall certainly, as before, obtain r non-zero elements $a_{11}, a_{22}, \dots, a_{rr}$, so that the leading minor of order r in the canonical form is $a_{11}a_{22} \dots a_{rr} \neq 0$. Since the rank of $H'AH$ is also r it follows that no element $a_{r+k,r+l}$ $(k, l > 0)$ can be non-zero. Hence the congruent canonical form of the matrix A is

$$H'AH = \begin{bmatrix} a_{11} & \cdot & \cdot & \cdots & \cdot & \cdots & \cdot \\ a_{21} & a_{22} & \cdot & \cdots & \cdot & \cdots & \cdot \\ a_{31} & a_{32} & a_{33} & \cdots & \cdot & \cdots & \cdot \\ \vdots & & & \cdot & & & \\ \vdots & & & & a_{rr} & & \cdot \\ \vdots & & & & \vdots & & \\ a_{n1} & \cdots & & \cdots & a_{nr} & & \cdot \end{bmatrix}, \quad . \quad . \quad . \quad (31)$$

or $H'AH = [a_{ij}]$, where $a_{ij} = 0$, $\begin{cases} i < j, \\ j > r. \end{cases}$

Thus the non-zero part of the canonical matrix appears as a truncated triangle, or trapezium. On introducing cogredient variables

x and y we obtain a bilinear form $y'Ax$ which gives rise to a correlation

$$\Gamma = y'Ax = \sum_{i,j=1}^{n} y_i a_{ij} x_j = 0. \quad . \quad . \quad . \quad . \quad (32)$$

By cogredient transformations in $\mathcal{F}$, namely $x = H\xi$, $y = H\eta$, the canonical form of Γ is obtained as

$$\Gamma = \eta' H' A H \xi = \sum_{i=1}^{n} \sum_{j=1}^{r} \eta_i a_{ij} \xi_j, \quad i > j. \quad . \quad . \quad (33)$$

10. **The Orthogonalizing Process of Schmidt.**

Consider a general matrix C in the field $\mathcal{F}$, and its transposed conjugate $\tilde{C}'$. If we write A for the product $\tilde{C}'C$ it follows that

$$\tilde{A} = \widetilde{(\tilde{C}'C)} = C'\tilde{C} = (\tilde{C}'C)' = A', \quad . \quad . \quad . \quad (34)$$

so that A is symmetric (or Hermitian) as before. If now we reduce A to the diagonal form $B = [a_i \delta_{ij}]$, we have

$$B = \tilde{H}'\tilde{C}'CH = (\widetilde{H'C'})CH = (\widetilde{CH})'(CH) = \tilde{D}'D, \quad . \quad (35)$$

where $D = CH$. This type of matrix D is of considerable importance, possessing as it does the property

$$\tilde{D}'D = \begin{bmatrix} a_1 & & & & \\ & a_2 & & & \\ & & \ddots & & \\ & & & a_r & \\ & & & & \end{bmatrix}. \quad . \quad . \quad . \quad . \quad (36)$$

If $r = n$ we shall say that D is *orthogonalized*; if $r < n$ we shall say that D is *semi-orthogonalized*, or *orthogonalized of rank* r. If we write $D = [d_1, d_2, \ldots, d_n]$, where d_j denotes the jth column of D, then $\tilde{D}' = \{\tilde{d}_1', \tilde{d}_2', \ldots, \tilde{d}_n'\}$, where $\tilde{d}_i'$ denotes the ith row of $\tilde{D}'$; and the identity (35) may be written in the form of orthogonal conditions

$$\left.\begin{aligned} (\tilde{d}_i \mid d_j) &= \sum_{k=1}^{n} \tilde{d}_{ki} d_{kj} = 0, \qquad i \neq j, \\ (\tilde{d}_i \mid d_i) &= \sum_{k=1}^{n} \tilde{d}_{ki} d_{ki} = a_i. \end{aligned}\right] \quad . \quad . \quad . \quad (37)$$

Briefly, $$\Sigma\, \tilde{d}_{ki}\, d_{kj} = a_i \delta_{ij}.$$

In the real field and also in the Hermitian case, each term $\tilde{d}_{ki}\, d_{ki}$

is the square of the modulus of d_{ki} and is therefore positive or zero. Hence $a_i > 0$, $i \leqslant r$, so that D is positive definite or non-negative definite.

If $r = n$ and $a_i = 1$, then $\tilde{D}'D = I$, and D is orthogonal or unitary according as $\tilde{D} = D$ or $\tilde{D} = \bar{D}$. If C is non-singular, then all $a_{ii} \neq 0$, for $a_{ii} = \sum_j c_{ji} c_{ji}$, a sum which cannot vanish unless each element c_{ji} of the ith column of C is zero. Hence the reduction (i) of Theorem I is unnecessary in forming B: consequently H is a continued product of matrices always with zeros below the principal diagonal, and is therefore triangular. This leads to an interesting theorem which, though purely algebraic, was first given, it appears, in connexion with integral equations.

Schmidt's Theorem.—*In the complex field there exists a non-singular matrix of triangular shape,* $K = [k_{ij}]$, $(k_{ij} = 0,\ i < j)$ *which transforms any given non-singular matrix of order* n *by post-multiplication into a unitary matrix.*

Proof.—From the given matrix C construct $D = CH$ as above. Since C is non-singular its rank is n and therefore $a_n \neq 0$, and D is orthogonalized. Postmultiply both D and H by the non-singular diagonal matrix D_1 having for (ii)th element $1/|a_i|^{\frac{1}{2}}$. This *normalizes* the ith column of D and also of $\tilde{D}$: further $\tilde{D}_1'\tilde{D}'DD_1 = I$. Hence DD_1 is unitary and the required matrix K is the matrix HD_1, which is triangular. This proves the theorem.

Corollary.—*Within the real field such a transformation turns a non-singular matrix by postmultiplication into an orthogonal matrix.*

11. Observations on Schmidt's Theorem.

Schmidt's Theorem can be regarded as an operation on a set of n vectors. If $x, y, \ldots, z$ are the n vectors (of the second kind) constituting D, so that

$$D = [x, y, \ldots, z] = \begin{bmatrix} x_1 & y_1 & \cdots & z_1 \\ x_2 & y_2 & \cdots & z_2 \\ \vdots & \vdots & \cdots & \vdots \\ x_n & y_n & \cdots & z_n \end{bmatrix},$$

and if $\tilde{D}'D = I$, then we have

$$\left.\begin{aligned} \tilde{x}'x = \tilde{y}'y = \ldots = \tilde{z}'z = 1, \\ \tilde{x}'y = \ldots = 0. \end{aligned}\right] \quad . \quad . \quad . \quad (38)$$

The vectors are normalized and orthogonal (or unitary) in pairs. Thus what Schmidt's process does is to take any n linearly independent vectors $\alpha, \beta, \ldots, \gamma$ and strain them, as it were, into an orthogonal or unitary set by a triangular matrix K, so that

$$[\alpha, \beta, \ldots, \gamma] \,.\, K = [x, y, \ldots, z]. \quad . \quad . \quad . \quad (39)$$

It is an example of n cogredient transformations: the n points $\alpha, \beta, \ldots, \gamma$ are strained into orthogonal positions, much as three conjugate diameters of a real quadric may be strained to principal axes by real transformations. In the complex field Schmidt's Theorem generalizes to the complex variable the transformation from oblique to rectangular axes.

It may be observed that the number of orthogonal conditions in (37) above is $\frac{1}{2}n(n-1)$, which is also the number of degrees of freedom in the choice of a simplex of reference, such as the triangle at the beginning of this chapter. The presumption is that there may exist an *orthogonal* or *unitary* matrix, which transforms any generalized quadratic form $\bar{x}'Ax$ into canonical shape: and this leads us to the more specialized transformations of the next Chapter.

EXAMPLES

1. If A is a non-singular matrix with complex elements, then the Hermitian matrix $\bar{A}'A$ is positive definite.

[For the Hermitian form $\bar{x}'\bar{A}'Ax$ equals $\bar{y}'y$, a sum of squared moduli of elements of the vector $y = Ax$. Since A is non-singular, y cannot vanish except in the excluded trivial case $x = 0$. Hence the Hermitian form is positive and not zero for every non-trivial value of x, and so must be positive definite and of full rank. Hence the matrix $\bar{A}'A$ is positive definite.]

2. If A is a non-singular real matrix, the real symmetric matrix $A'A$ is positive definite.

3. If A is a rectangular matrix of order $m \times n$, with columns of complex elements, the Hermitian matrix $\bar{A}'A$ is positive definite if $m > n$ and the columns of A are linearly independent, and non-negative definite if $m < n$.

4. *The determinant of a positive definite Hermitian matrix* A *cannot exceed in value the continued product of the diagonal elements* a_{ii}.

[Consider the first stage of the reduction at (11), p. 85. The matrix $\bar{b}'b$ has non-negative diagonal elements $\bar{a}_{1j}a_{1j}$, and a_{11} is positive. Hence the leading element of $A_1 - \bar{b}'a_{11}^{-1}b$ is $\alpha_2 \leqslant a_{22}$. Since A is positive definite $\alpha_2 > 0$, and $a_{22} > 0$. Thus at the next stage we shall likewise have $\alpha_3 \leqslant a_{33}$, and so on. Also all the transformations used are unimodular. Hence

$$|A| = \alpha_1\alpha_2 \ldots \alpha_n \leqslant a_{11}a_{22} \ldots a_{nn}.]$$

5. *The modulus of the determinant of a general complex matrix* A *cannot exceed*

$$(\Pi\bar{a}'a)^{\frac{1}{2}},$$

where a *denotes a column of* A, *and the product is taken over all columns.* (Hadamard's theorem.)

[The singular case being trivial, we take A to be non-singular. Then, as in Ex. 1 above, the Hermitian matrix $\bar{A}'A$ is positive definite, its diagonal elements being the real scalars $\bar{a}'a$. The result now follows at once from Ex. 4.]

6. The determinant of a positive definite Hermitian matrix A cannot exceed $a_{11} | A_{11} |$, where $| A_{11} |$ is the co-factor $| a_{22}a_{33} \ldots a_{nn} |$ of a_{11}.

[By conjunctive operations $\text{row}_i - k\ \text{row}_2$, $\text{col}_i - k\ \text{col}_2$ we may remove all elements in the first row and column except a_{11}, a_{12} and $a_{21} = \bar{a}_{21}$, without affecting the positive definite character of A_{11}, or altering the value of $| A_{11} |$. Expanding $| A |$ in terms of its first row and column, we have

$$| A | = a_{11} | A_{11} | - \bar{a}_{12}a_{12} | A_{12,\ 12} |,$$

where $| A_{12,\ 12} |$ denotes the co-factor of $a_{11}a_{22}$. This co-factor is non-negative, for the reason given above, as also is its coefficient $\bar{a}_{12}a_{12}$. The result now follows, and could be used to establish the theorem of Ex. 4.]

7. The determinant of a positive definite Hermitian matrix A cannot exceed the product of two complementary principal minors.

[Take the minors of orders m and $n - m$, and consider the argument of Ex. 6 in relation to the positive definite Hermitian compound matrix $A^{(m)}$.]

8. The determinant of a positive definite Hermitian matrix cannot exceed the continued product of any complete set of non-overlapping principal minors.

9. Enunciate the results of the preceding examples for the case of a positive definite real symmetric matrix.

10. If a denote a column of a general matrix A, and u denote the corresponding *row* of the adjugate of A, then $\bar{a}'a \,.\, u\bar{u}' \geqslant | \bar{A} | | A |$.

[Let M denote the matrix of two columns $[a, \bar{u}']$. The matrix

$$\bar{M}'M = \begin{bmatrix} \bar{a}'a & \bar{a}'\bar{u}' \\ ua & u\bar{u}' \end{bmatrix} = \begin{bmatrix} \bar{a}'a & | \bar{A} | \\ | A | & u\bar{u}' \end{bmatrix}$$

is then a non-negative definite matrix of the second order, and so

$$\bar{a}'a \,.\, u\bar{u}' \geqslant | \bar{A} | \ | A |.]$$

11. If each of the elements of a complex matrix A is of unit modulus, the modulus of the determinant of A cannot exceed $n^{n/2}$.

[Apply Hadamard's theorem, Ex. 5.]

12. If $A = -\bar{A}'$ is a skew Hermitian matrix, then $B = iA$ is an Hermitian matrix.

12. **Historical Note.**—The transformation of a quadratic form into a sum of squares dates back to Lagrange (1759); *Œuvres*, I, 3–20. It was also carried out by Gauss (1823), *Werke*, IV, 27–54; later by Jacobi, Brioschi, Kronecker, and many others. The concept of rank

was explicitly discussed by Sylvester, *Phil. Mag.* (4), **1** (1851), p. 121, or *Coll. Papers*, I, 221; he gave the law of inertia in the following year; *Phil. Mag.* (4), **4** (1852), p. 142, or *Coll. Papers*, I, 380. Jacobi gave it in *J. für Math.*, **53** (1857), 275. The determinantal criteria for rank and definiteness are due to Frobenius, *J. für Math.*, **82** (1876), 241–5, *Berl. Sitzungsb.* (1894), p. 245. Schmidt's Theorem is found in *Math. Annalen*, **63** (1907), 442; see also Schur, *Math. Zeits.*, **1** (1918), 205.

The chapters on the history of axisymmetric determinants in Muir's *History* contain many results bearing on quadratic forms. The chapters on skew determinants are also relevant.

Concerning Hadamard's theorem on the upper bound of the modulus of a determinant, we may refer to Muir's *History*, 1900–20, Chapter I(a), and in particular to the contributions of Fischer, Szász, and Cipolla. The theorem is of date 1893, but Muir, at the suggestion of Lord Kelvin, had investigated it in 1885, and later did so in more detail in 1908. The geometrical meaning of the theorem is that a parallelepiped of given length of edges has maximum volume when those edges are mutually perpendicular.

CHAPTER VIII

Canonical Reduction by Unitary and Orthogonal Transformations

As has just been seen, the reduction of a quadratic or Hermitian form to a sum of squares or of squared moduli is far from unique, considerable scope being left for the choice of the matrix H in the transformation

$$\tilde{H}'AH = [a_i \delta_{ij}] = D. \quad . \quad . \quad . \quad . \quad . \quad (1)$$

It is natural to inquire whether amongst these conjunctive reductions there exist any which are also collineatory; any such, by Chapter IV, would be unitary, or orthogonal, with $\tilde{H}'H = I$. Then too not only the matrices A and D but also the families, or characteristic matrices, $A - \lambda I$, $D - \lambda I$, would be equivalent, for

$$\tilde{H}'(A - \lambda I)H = \tilde{H}'AH - \lambda \tilde{H}'H = D - \lambda I. \quad . \quad . \quad (2)$$

It is indeed the possession of the double advantage of congruent and collineatory properties that gives the orthogonal and unitary sub-groups their peculiar importance. It is this that underlies the reduction of conics and quadric surfaces to principal axes, and the extensions in analysis to orthogonal functions.

1. The Latent Roots of Hermitian and Real Symmetric Matrices.

In preparation for what is to follow we shall next consider the nature of the latent roots of Hermitian matrices. It is a well-known feature in the reduction of conics and quadrics to principal axes that the roots of the discriminating quadratic or cubic $| a_{ij} - \lambda \delta_{ij} | = 0$ are real. This is only a special case of a property possessed by generalized symmetric matrices $[a_{ij}]$, where $a_{ji} = \tilde{a}_{ij}$.

Example:

$$A = \begin{bmatrix} 3 & 2i \\ -2i & . \end{bmatrix} = \begin{bmatrix} 3 & . \\ . & . \end{bmatrix} + \begin{bmatrix} . & 2i \\ -2i & . \end{bmatrix} = Q + iS. \quad (3)$$

The latent roots λ of the above matrix are given by $\lambda^2 - 3\lambda - 4 = 0$. Thus $\lambda = 4, -1$. The corresponding latent points are given by

$3x_1 + 2ix_2 = \lambda x_1$, $-2ix_1 = \lambda x_2$. They are therefore $\{2i, 1\}$ and $\{1, 2i\}$, or multiples of these, respectively. It is to be observed that the latent *points* of Hermitian matrices are not necessarily real.

Theorem I.—*The latent roots of an Hermitian matrix are real.*

Proof.—Let $A = \bar{A}'$ be the matrix, and let λ be a latent root corresponding to the latent point z. Then

$$Az = \lambda z, \quad z \neq 0, \quad \ldots \ldots \quad (4)$$

and so

$$\bar{z}'Az = \lambda \bar{z}'z. \quad \ldots \ldots \ldots \quad (5)$$

But (p. 84) the Hermitian forms $\bar{z}'Az$ and $\bar{z}'z$ are real, and $\bar{z}'z$ non-zero: hence λ is real, which proves the theorem.

Corollary I.—*The latent roots of a real symmetric matrix are real.*

Corollary II.—*The latent roots of a real skew symmetric matrix are either zero or else pure imaginaries, conjugate in pairs. If the matrix is of odd order, one at least of the latent roots is zero.*

The first corollary is obtained (cf. p. 34) by putting $S = 0$ in Theorem I: the second by putting $Q = 0$, for then the latent roots of the matrix iS are real, so that we have for example $iSz = \lambda z$. Hence $Sz = -i\lambda z$ and the roots of S are pure imaginary.

EXAMPLES

1. Prove that if the n latent roots of a real symmetric matrix are distinct, they are in general separated by the $n - 1$ latent roots of each diagonal submatrix of order $n - 1$.

[Consider the real symmetric matrix and its adjoint matrix

$$Q \equiv \begin{bmatrix} a & h & g \\ h & b & f \\ g & f & c \end{bmatrix}, \qquad \text{adj } Q \equiv \begin{bmatrix} A & H & G \\ H & B & F \\ G & F & C \end{bmatrix},$$

where $A = bc - f^2$, &c. The latent roots γ_1, γ_2 of C are given by

$$C_\lambda \equiv (a - \lambda)(b - \lambda) - h^2 = 0.$$

By putting $\lambda = -\infty$, a (or b), $+\infty$ we infer that both a and b lie between γ_1 and γ_2 when $h \neq 0$. Thus $\gamma_1 < a < \gamma_2$; $\gamma_1 < b < \gamma_2$.

Again, by Jacobi's theorem on adjugate determinants (*Invariants*, p. 77), $B_\lambda C_\lambda - F_\lambda{}^2 = (a - \lambda)\Delta_\lambda$, where Δ_λ denotes $|Q - \lambda I|$. *In general* we have $F_\lambda \neq 0$; hence at γ_1, γ_2 the value of $(a - \lambda)\Delta_\lambda$ is negative. By putting $\lambda = -\infty, \gamma_1, \gamma_2, +\infty$ we find that the signs of Δ_λ are $+, -, +, -$. Hence $\lambda_1 < \gamma_1 < \lambda_2 < \gamma_2 < \lambda_3$, where the λ_i are the latent roots of Q. If we apply Jacobi's theorem for diagonal submatrices of order $n - 1$, the proof follows by induction.

If each non-diagonal minor (such as H_λ) is non-zero at a latent root of Q, the general theorem is true. But, if $H_\lambda = 0$, then $\Delta_\lambda = 0$. Simultaneously, a latent root of Q coincides with one of H. The reader should also consider the cases (i) when Q is a diagonal matrix, and (ii) when Q has repeated latent roots.]

2. Investigate the corresponding theorem for Hermitian matrices.

3. A continuant matrix will be defined as one which has zero elements except in its diagonal, its first superdiagonal, and its first subdiagonal; $a_{i,\, i\pm r} = 0$, $r > 1$. Prove that if $a_{i,\, i+1}$ and $a_{i+1,\, i}$ are real and of the same sign for each i, the latent roots of a continuant matrix are real.

[A collineatory transformation $\{H, H^{-1}\}$ can be found involving a real diagonal matrix $H = [h_i \delta_{ij}]$, which reduces the continuant to an axisymmetric matrix.]

4. The Hermite, Legendre, Tchebychef and other orthogonal polynomials of analysis can be expressed as characteristic determinants of continuants of the above type by solving from their recurrence relations. Hence the roots of these polynomials are all real.

[For example the fourth Hermite polynomial is $\begin{vmatrix} x & 1 & \cdot & \cdot \\ 1 & x & 2 & \cdot \\ \cdot & 1 & x & 3 \\ \cdot & \cdot & 1 & x \end{vmatrix} = x^4 - 6x^2 + 3,$

and $\begin{bmatrix} \cdot & -1 & \cdot & \cdot \\ -1 & \cdot & -2 & \cdot \\ \cdot & -1 & \cdot & -3 \\ \cdot & \cdot & -1 & \cdot \end{bmatrix}$ can be transformed to $\begin{bmatrix} \cdot & 1 & \cdot & \cdot \\ 1 & \cdot & \sqrt{2} & \cdot \\ \cdot & \sqrt{2} & \cdot & \sqrt{3} \\ \cdot & \cdot & \sqrt{3} & \cdot \end{bmatrix}$ by $[h_i \delta_{ij}]$,

where $h_1 = -1/\sqrt{6}$, $h_2 = 1/\sqrt{6}$, $h_3 = -1/\sqrt{3}$, $h_4 = 1$.]

5. Transform to axisymmetric type the continuant matrix corresponding to the fourth Legendre polynomial, namely $\begin{bmatrix} \cdot & 1 & \cdot & \cdot \\ \frac{1}{3} & \cdot & \frac{2}{3} & \cdot \\ \cdot & \frac{2}{5} & \cdot & \frac{3}{5} \\ \cdot & \cdot & \frac{3}{7} & \cdot \end{bmatrix}$.

6. If in a continuant matrix $a_{i,\, i} = 0$ and $a_{i,\, i+1}$ and $a_{i+1,\, i}$ are of *opposite* sign, the latent roots are pure imaginary if the matrix is of even order, together with one zero root if the matrix is of odd order.

2. The Concept of Rotation Generalized.

Before we deal with the orthogonal or unitary transformation of matrices we must consider such transformations for column vectors. Here the inequality

$$\bar{z}'z > 0, \qquad (6)$$

although its meaning is simply that the norm $\bar{z}'z$ of the latent point z is positive, proves indispensable to further progress. An example may serve to show how it may fail to hold in the case of a complex *symmetric* matrix possessing a latent point p of norm $p'p$.

Consider the complex symmetric matrix $A = A' = \begin{bmatrix} 2 & i \\ i & . \end{bmatrix}$, which has latent roots 1, 1. If p is a corresponding latent point, we have $2p_1 + ip_2 = p_1$, $ip_1 = p_2$. Hence $p = \{1, i\}$ to a constant factor, and $p'p = 0$; so that the inequality in question does not hold. The vector p is in fact *isotropic* and cannot be normalized; and this, as we shall see, renders it useless for our purpose. We therefore confine the discussion to the Hermitian and the real symmetric cases.

It is a commonplace of kinematics that by a rotation about a suitable axis through an origin O a line OP may be brought into coincidence with a line OQ of equal length. Algebraically this concerns two vectors p, q with equal norms $p'p$ and $q'q$; and it means that an orthogonal matrix R exists such that $Rp = q$. We shall extend this to the case of n dimensions and an Hermitian matrix.

Lemma I.—*From an arbitrary non-zero column vector of complex elements a matrix may be constructed which shall at the same time be rational in the elements, Hermitian, unitary, and involutory.*

Proof.—From z, the chosen vector, construct $\theta = \bar{z}'z$, $Z = z\bar{z}'$. As was found in Chapter I, p. 4, θ is scalar, but Z is a square matrix of order n. Since $z \neq 0$, θ is the norm of z and is positive: on the other hand, Z is Hermitian. Thus

$$\theta = \bar{z}'z = z'\bar{z} > 0; \quad Z = z\bar{z}' = (\overline{z\bar{z}'})' = \bar{Z}'. \quad . \quad . \quad . \quad (7)$$

Also
$$Z^2 = z\bar{z}'z\bar{z}' = z\theta\bar{z}' = \theta z\bar{z}' = \theta Z,$$

so that
$$Z^2 - \theta Z = \bar{Z}'Z - \theta Z = 0. \quad . \quad . \quad . \quad . \quad . \quad . \quad (8)$$

It is noted in passing that the R.C.F. of Z is a quadratic, so that each of the n latent roots of Z is either zero, occurring $n - 1$ times, or θ Again, by (8),

$$\left(\frac{2}{\theta}\bar{Z}' - I\right)\left(\frac{2}{\theta}Z - I\right) = I, \quad \theta > 0. \quad . \quad . \quad . \quad . \quad (9)$$

A matrix $Q = \frac{2}{\theta}Z - I$ has thus been rationally constructed which has the properties

$$Q = \bar{Q}', \quad \bar{Q}'Q = I, \quad Q^2 = I, \quad . \quad . \quad . \quad (10)$$

and which is therefore Hermitian, unitary, and involutory.

(*Note.*—If Q transforms a point x to ξ, so that $Qx = \xi$, then

$Q \cdot \xi = Q^2 x = x$. Thus in involutory transformations object and image points are interchangeable.)

The latent points of the above unitary matrix Q are z and multiples of z. For $Qz = 2\theta^{-1} Z z - z = 2z - z = z$. Hence z is a latent point.

It may make for clearness to interpret the above work in terms of rectangular Cartesian co-ordinates, regarding the vector $\alpha z = \{\alpha z_1, \alpha z_2, \ldots, \alpha z_n\}$ as a line of points through the origin O in space of n dimensions. Evidently this line, Oz say, is pointwise latent: the transformation Q leaves each point unchanged. If we take another point x and transform it to ξ by $Qx = \xi$, we find at once that

$$Q(x + \xi) = \xi + x. \quad \ldots \ldots \quad (11)$$

This shows that the *middle* point of the line joining ξ to x is latent for all positions of x. We infer that if $n = 3$, the transformation represents a rotation through two right angles about an axis through O and the point z: and that, for $n > 3$ and for the complex case, the algebra generalizes this concept of rotation through two right angles. We leave it to the reader to consider the real and the complex cases when $n = 2$, with the remark that when $n = 1$ the complex case is a particular instance of the unitary transformation $Qx = \xi$, $Q = e^{i\alpha}$ where $\alpha = \pi$. It is a fundamental fact that the norm of a vector is invariant under unitary transformation; thus the distances from the origin of the points x and ξ above are equal.

Lemma II.—*Any two non-zero vectors of order* n *having equal norms may be transformed into each other by unitary transformations.*

Proof.—Let p and q be the vectors. Using the converse of Lemma I, we take z to be any point except O on the line through the point $p + q$. Then we have, supposing the vectors normalized,

$$z = p + q, \quad \theta = (\bar{p}' + \bar{q}')(p + q) = 2 + \rho + \bar{\rho},$$

where

$$\bar{p}'p = \bar{q}'q = 1, \quad \bar{p}'q = \bar{\rho}, \quad \bar{q}'p = \rho \quad \ldots \quad (12)$$

Then, if as before $Z = z\bar{z}'$, the matrices

$$R = \frac{Z}{1 + \rho} - I, \quad \bar{R} = \frac{\bar{Z}}{1 + \bar{\rho}} - I \quad \ldots \quad (13)$$

are unitary, since $\bar{Z}' = Z$, and

$$\bar{R}'R = \left(\frac{\bar{Z}'}{1 + \bar{\rho}} - I\right)\left(\frac{Z}{1 + \rho} - I\right) = \frac{Z^2 - Z\theta + (1 + \bar{\rho})(1 + \rho)I}{(1 + \bar{\rho})(1 + \rho)} = I.$$

Also $1 + \rho = (\bar{p}' + \bar{q}')p$. Hence

$$Zp = (p + q)(\bar{p}' + \bar{q}')\,p = (p + q)(1 + \rho),$$

or

$$\left(\frac{Z}{1 + \rho} - I\right) p = q,$$

i.e. $Rp = q$. Similarly $\bar{R}q = p$. . . . . (14)

Two conjugate unitary matrices R and $\bar{R}$ have thus been constructed which rationally transform vectors of equal norms, or unit vectors, p to q and q to p respectively. A case of failure arises if $1 + \rho$ vanishes. This is obviated by making the transformation R first from p to $-q$ (which is possible since $1 + \rho$ and $1 - \rho$ cannot both be zero), followed by a transformation from $-q$ to q by means of the negative matrix $-I$, which is unitary. This proves the Lemma.

Corollary.—*A unitary matrix* R *exists which transforms a given vector of norm* $\bar{\alpha}\alpha$ *into* $\{\alpha, 0, 0, \ldots, 0\}$.

Theorem II.—*A square matrix* A *of order* n *may be reduced to the canonical form of Jacobi,* $\Gamma = [\gamma_{ij}]$, $i \leqslant j$, *by a transformation* RAR^{-1}, *where* R *is a unitary matrix in the field of the elements and of the latent roots of* A.

Proof.—This is almost identical with the proof of the Jacobian reduction in the collineatory case, p. 65. By the Corollary above, a unitary matrix R_1 exists which transforms any given non-zero vector p into $q = \{\alpha, 0, 0, \ldots, 0\}$. Let p be a pole of A, so that $Ap = \lambda_1 p$. Then, as before, $R_1 A R_1^{-1}$ has for its first column $\{\lambda_1, 0, 0, \ldots, 0\}$. The rest of the proof follows exactly as in the former case. By the group property of unitary matrices the various transformations by R_1 may be combined into a single unitary transformation R, and we have

$$\Gamma = RAR^{-1} = \begin{bmatrix} \lambda_1 & \gamma_{12} & \cdots & \gamma_{1n} \\ & \lambda_2 & \cdots & \gamma_{2n} \\ & & \ddots & \vdots \\ & & & \lambda_n \end{bmatrix}, \quad \gamma_{ij} = 0, \quad i > j, \qquad (15)$$

where the diagonal elements are the latent roots of A arranged in any prescribed order.

Theorem III.—*If the matrix* A *is Hermitian, a unitary matrix* R *exists which transforms* A *into a diagonal canonical form of real elements* λ_i, *namely* $D = [\lambda_i \delta_{ij}]$.

Proof.—Since a unitary transformation preserves the Hermitian property, the matrix which corresponds to Γ in the present case must be Hermitian, with $\bar{\gamma}_{ij} = \gamma_{ji}$. Hence it must have zeros above the diagonal as well as below. Also its diagonal elements can only be the latent roots in some order. Hence the theorem is true.

This reduction gives an alternative demonstration of the fact that the latent roots of Hermitian and real symmetric matrices are real; for the canonical form is also Hermitian or real symmetric, so that the latent roots in the diagonal are equal to their conjugates and therefore real. The imaginary nature of the latent roots of skew matrices may be similarly proved.

It has thus been proved that by a unitary transformation we may reduce an Hermitian form $\bar{x}'A\,x$ to the canonical shape

$$\bar{x}'Ax = \bar{\xi}'D\xi = \lambda_1\bar{\xi}_1\xi_1 + \lambda_2\bar{\xi}_2\xi_2 + \ldots + \lambda_r\bar{\xi}_r\xi_r, \qquad (16)$$

where $Rx = \xi$, $\bar{R}'R = I$, and $RAR^{-1} = D$.

The reduction is no longer necessarily rational, but lies in the field of all complex numbers in the Hermitian and unitary case, and in the field of all real numbers in the real symmetric and orthogonal case.

3. The Canonical Reduction of Pairs of Forms or Matrices.

The above results can be formulated in another way. Just as a rotation leaves certain distances invariant, so an orthogonal transformation leaves $x'x$ invariant and a unitary transformation leaves $\bar{x}'x$ invariant. But $x'x$ and $\bar{x}'x$ are quadratic forms $x'Ix$ and $\bar{x}'Ix$ with unit matrix I, and we have also $\bar{R}'IR = I$. It is natural to express this by saying that the matrix I is *latent* in the unitary transformation R; and more generally if $PAQ = A$ we shall say that the matrix A is latent in the transformation (P, Q). Conversely, a conjunctive or congruent transformation which leaves the unit matrix latent can only be unitary or orthogonal, as is readily verified. (Cf. Chapter IV, p. 39; or *Invariants*, p. 152.) Now let there be a pair of Hermitian forms

$$f = \bar{x}'A\,x, \quad \phi = \bar{x}'B\,x, \quad \bar{A}' = A, \; \bar{B}' = B, \quad . \quad . \quad (17)$$

such that A is of rank r and B is positive definite. By a rational transformation $x = P\xi$, $\bar{x} = \bar{P}\bar{\xi}$ we can reduce ϕ to the form $\bar{\xi}'[\beta_i\delta_{ij}]\xi$ as above, where β_i is real and positive. Simultaneously A becomes $\bar{P}'AP$. By a real diagonal transformation the n real non-zero elements β_1 can become units, so that the matrix B is now transformed to I, the unit matrix. At the same time let $\bar{P}'AP$ become $\bar{Q}'AQ$. By a unitary transformation we can reduce $\bar{Q}'AQ$ to diagonal form $[a_i\delta_{ij}]$. Combining the three transformations, which are all conjunctive, we see that there exists a conjunctive transformation H which simultaneously reduces A and B to diagonal forms

$$\bar{H}'AH = \begin{bmatrix} a_1 & & & & & & \\ & a_2 & & & & & \\ & & \cdot & & & & \\ & & & \cdot & & & \\ & & & & a_r & & \\ & & & & & \cdot & \\ & & & & & & \cdot \end{bmatrix}, \quad \bar{H}'BH = I. \quad . \quad . \quad (18)$$

All of the r coefficients a_i are real and non-zero.

Thus, in the real field, two real quadratic forms, one of which has rank r and the other of which is positive definite, can be reduced simultaneously to the sums of squares

$$\left. \begin{aligned} a_1y_1^2 + a_2y_2^2 + \ldots + a_ry_r^2, \\ y_1^2 + y_2^2 + \ldots + y_n^2. \end{aligned} \right\} \quad . \quad . \quad . \quad (19)$$

The discovery of the transformation H contains the solution of the problem of principal axes for conics and quadrics, in non-Euclidean geometry as well as Euclidean.

The following examples illustrate a number of interesting and more or less isolated unitary and orthogonal properties.

EXAMPLES

1. Prove that if Q is a real symmetric matrix and S is a real skew symmetric matrix of the same order, then the matrix $X = (I + S + iQ)(I - S - iQ)^{-1}$ is unitary. (Cf. *Invariants*, p. 156.) [Note that the given factors of X are permutable.]

2. If $Q = \begin{bmatrix} a & h \\ h & b \end{bmatrix}$, $S = \begin{bmatrix} \cdot & s \\ -s & \cdot \end{bmatrix}$, construct X. Construct also the corresponding unitary matrix for the third order.

3. The latent roots of a unitary matrix are all of unit modulus.

[Let R be a unitary matrix, λ a latent root, z the latent point corresponding. Then $Rz = \lambda z$, $\bar{z}'\bar{R}' = \bar{\lambda}\bar{z}'$, and so $\bar{z}'\bar{R}'Rz = \bar{z}'Iz = \bar{z}'z = \bar{\lambda}\lambda\bar{z}'z$. Since $z \neq 0$, $\bar{\lambda}\lambda = 1$.]

4. Prove that if a, b, c are complex scalars, then the matrix (Hermitian or symmetric)

$$\begin{bmatrix} \dfrac{\bar{a}a - \bar{b}b - \bar{c}c}{\bar{a}a + \bar{b}b + \bar{c}c} & \dfrac{\bar{a}b}{\bar{a}a + \bar{b}b + \bar{c}c} & \dfrac{\bar{a}c}{\bar{a}a + \bar{b}b + \bar{c}c} \\ \dfrac{\bar{b}a}{\bar{a}a + \bar{b}b + \bar{c}c} & \dfrac{\bar{b}b - \bar{c}c - \bar{a}a}{\bar{a}a + \bar{b}b + \bar{c}c} & \dfrac{\bar{b}c}{\bar{a}a + \bar{b}b + \bar{c}c} \\ \dfrac{\bar{c}a}{\bar{a}a + \bar{b}b + \bar{c}c} & \dfrac{\bar{c}b}{\bar{a}a + \bar{b}b + \bar{c}c} & \dfrac{\bar{c}c - \bar{b}b - \bar{a}a}{\bar{a}a + \bar{b}b + \bar{c}c} \end{bmatrix}$$

is unitary, or orthogonal. [Use Lemma I.]

5. A real skew symmetric matrix S may be transformed by a unitary transformation into a diagonal form with pure imaginary elements.

[Put $Q = 0$ in the Hermitian case and reduce iS to $[\alpha_i \delta_{ij}]$, where α_i is real.]

6. The roots of the secular equation $| A - \lambda B | \equiv | a_{ij} - \lambda b_{ij} | = 0$ are real if $\bar{A}' = A$, $\bar{B}' = B$, and B is positive definite. The elementary divisors of the λ matrix $A - \lambda B$ are *linear*.

[Take a non-zero latent point $x + iy$ such that $(A - \lambda B)(x + iy) = 0$. Proceed as in Theorem I. Use (18).]

7. If both A and B above are positive definite, the roots of $| A - \lambda B | = 0$ have one sign.

8. Prove that $\begin{vmatrix} x & a \\ -a & x \end{vmatrix} > 0$, $\begin{vmatrix} x & a & b \\ -a & x & c \\ -b & -c & x \end{vmatrix} >, =, < 0$, if $x >, =, < 0$, and so on, where the elements of the matrices are all real.

9. Through the curve of intersection of an ellipsoid $\Sigma = 0$ and another quadric surface $\Sigma' = 0$, three real paraboloids can be drawn. (Hilton.)

10. An ellipsoid and a concentric quadric have in common a real trio of conjugate diameters. (Hilton.)

11. A unitary matrix A can be transformed by another unitary matrix R into diagonal form.

[Since $\bar{A}'A = I$ the rank of A is n, and so none of the latent roots is zero. If z is a latent point of A, then $Az = \lambda z$, $z \neq 0$, $\lambda \neq 0$. Use a matrix R_1 such that $R_1 z = \{\alpha, 0, 0, \ldots, 0\}$ and proceed as in Theorems II and III.]

12. Use the result of Ex. 11 to prove that the latent roots of a unitary matrix have unit modulus.

[If A is reduced to $[\alpha_i \delta_{ij}]$, which is again unitary, then

$$[\bar{\alpha}_i \delta_{ij}]' \, [\alpha_i \delta_{ij}] = I = [\delta_{ij}]. \text{ Hence } \bar{\alpha}_i \alpha_i = 1.]$$

13. To construct a real orthogonal (or unitary) matrix whose first m columns are prescribed real orthogonal (or unitary) vectors.

[For example, take $m = 2$, $x'x = y'y = 1$, $x'y = 0$, where x, y are the given real column vectors. By the corollary of Lemma II there exists a real orthogonal matrix (cf. *Invariants*, p. 317) such that

$$A_1 x = \{1, 0, 0, \ldots, 0\} = \xi, \quad A_1' A_1 = I.$$

Let $A_1 y = \eta$. Then $\xi' \eta = x' A_1' A_1 y = x'y = 0$. On substituting for ξ, we have

$\eta = \{0, \eta_2, \eta_3, \ldots, \eta_n\}$, $\eta'\eta = y'y = 1$. Next a real orthogonal submatrix A_2 exists such that

$$A_2\{\eta_2, \ldots, \eta_n\} = \{1, 0, \ldots, 0\}.$$

Hence $\begin{bmatrix} 1 & \cdot \\ \cdot & A_2 \end{bmatrix} A_1 = A$ is an orthogonal matrix such that $A[x, y] = \begin{bmatrix} 1 & \cdot \\ \cdot & 1 \\ \cdot & \cdot \\ \cdot & \cdot \\ \vdots & \vdots \end{bmatrix}$.

It follows that A^{-1} (which is orthogonal and therefore non-singular) has, for its first two columns, x and y. This is the required matrix; and the mode of its formation applies for any number of columns, and also for unitary vectors.]

14. A real orthogonal matrix A can be reduced by a real orthogonal matrix P to a diagonal of submatrices,

$$PAP^{-1} = \begin{bmatrix} A_1 & & & & & \\ & A_2 & & & & \\ & & \cdot & & & \\ & & & \cdot & & \\ & & & & \cdot & \\ & & & & & A_k \end{bmatrix},$$

where each A_i is either 1, or -1, or $\begin{bmatrix} \cos\varphi & \sin\varphi \\ -\sin\varphi & \cos\varphi \end{bmatrix} \equiv \Phi$.

[By Examples 3, or 12, the latent roots are 1, or -1, or $e^{\pm i\phi}$ in pairs. Proceeding as in Ex. 11, we isolate a diagonal element ± 1, for each latent root ± 1, by means of a real *orthogonal* matrix R_1. The outstanding submatrix A has latent roots $e^{\pm i\phi}$ in pairs, and is therefore of even order. Take a pole $z = x + iy$, such that

$$Az = e^{i\phi}z, \qquad A'A = I.$$

Hence, separating real and imaginary parts,

$$Ax = x\cos\varphi - y\sin\varphi, \qquad Ay = x\sin\varphi + y\cos\varphi,$$

(where x and y are real column-vectors). Write $x'x = \rho$, $x'y = y'x = \sigma$, $y'y = \tau$, all of which are scalar. Then, by transposition,

$$x'A' = x'\cos\varphi - y'\sin\varphi, \qquad y'A' = x'\sin\varphi + y'\cos\varphi,$$

and by multiplication, (using $x'A' \,.\, Ax = x'x = \rho$),

$$\rho = \rho\cos^2\varphi - 2\sigma\cos\varphi\sin\varphi + \tau\sin^2\varphi.$$

Similarly, ($y'A' \,.\, Ax = y'x = \sigma$),

$$\sigma = \rho\cos\varphi\sin\varphi + \sigma\cos 2\varphi - \tau\cos\varphi\sin\varphi.$$

But $\sin\varphi \neq 0$. Hence $\rho = \tau$, $\sigma = 0$: or

$$x'x = y'y, \qquad x'y = 0.$$

Hence x and y are orthogonal vectors, and without loss of generality can be normalized, with $x'x = y'y = 1$. Again the above relations for Ax, Ay can be written

$$A[x, y] = [x, y]\begin{bmatrix} \cos\varphi & \sin\varphi \\ -\sin\varphi & \cos\varphi \end{bmatrix} = [x, y]\Phi;$$

and the requisite Φ matrix can now be semi-isolated by using Ex. 13. Thus if

$$B[x, y] = \begin{bmatrix} 1 & \cdot \\ \cdot & 1 \\ \cdot & \cdot \\ \cdot\ \cdot & \cdot\ \cdot \end{bmatrix}, \text{ where } B \text{ is orthogonal, we find that } BAB^{-1} = \begin{bmatrix} \Phi & C \\ \cdot & D \end{bmatrix}.$$

But BAB^{-1} is orthogonal: hence

$$\begin{bmatrix} \Phi' & \cdot \\ C' & D' \end{bmatrix} \begin{bmatrix} \Phi & C \\ \cdot & D \end{bmatrix} = I, \text{ i.e. } \Phi'\Phi = I,\ \Phi'C = 0,\ C'C + D'D = I.$$

Thus $0 = \Phi\Phi'C = C$, and $D'D = I$, showing that Φ is isolated and that the outstanding D is orthogonal. Repetition of the process, upon D, completes the proof of the theorem.

The determinant of the orthogonal matrix is $(-1)^k$, where k is the number of latent roots equal to -1. The matrix is a *proper* or *improper* orthogonal matrix according as k is even or odd.

If k is even, the isolated elements -1 may be grouped in pairs, and regarded as of type Φ with $\varphi = \pi$. If k is odd, a single isolated -1 is outstanding. This shows that an orthogonal transformation, in n-space, of rectangular Cartesian co-ordinates, with fixed origin, is built up of at most $\frac{1}{2}n$ rotations through suitable angles φ, supplemented (in the *improper* case) by the reversal of *one* axis.]

15. If $\mu + i\nu = \lambda$ is a latent root of a general matrix A of complex elements, then μ lies between the greatest and least latent root of the Hermitian matrix $P = \frac{1}{2}(A + \bar{A}')$, and ν lies between the greatest and least latent root of the Hermitian matrix $Q = \frac{1}{2}i(A - \bar{A}')$. (Bendixson, Hirsch, Bromwich.)

[The latent roots of P and Q are real. Let $z \neq 0$ be a pole of A with respect to λ. Then $Az = \lambda z$, $\bar{z}'\bar{A}' = \bar{\lambda}z'$. Hence $\bar{z}'Pz = \frac{1}{2}\bar{z}'(A + \bar{A}')z = \mu\bar{z}'z$. Reduce P to real diagonal form $[\alpha_i\delta_{ij}]$ by a unitary transformation R; μI is left unchanged. If $Rz = y$, we have then $\bar{y}'[\alpha_i\delta_{ij}]y = \mu\bar{y}'y$, that is,

$$\mu = (\Sigma\alpha_i\xi_i^2) / (\Sigma\, \xi_i^2), \text{ say.}$$

This shows that μ must lie between the greatest and the least of these real roots α_i.

Again, $\bar{z}'Qz = \frac{1}{2}i\bar{z}'(A - \bar{A}')z = \nu\bar{z}'z$, and a similar argument applies.]

16. If λ is a latent root of the complex matrix A, then $\bar{\lambda}\lambda$ lies between the greatest and the least latent roots of the non-negative definite Hermitian matrix $\bar{A}'A$. (E. T. Browne.)

[If $Az = \lambda z$, then $\bar{z}'\bar{A}'Az = \bar{\lambda}\lambda\bar{z}'z$. Transform $\bar{A}'A$, which is Hermitian, by a unitary transformation to a real non-negative diagonal $[\rho_i\delta_{ij}]$. If z becomes y, we have $\Sigma\rho_i\bar{y}_iy_i = \bar{\lambda}\lambda\Sigma\bar{y}_iy_i$, whence the result follows as in the previous example.

That $\bar{A}'A$ is non-negative definite follows from Ex. 3, p. 97.]

17. By taking the mth compound identity of that used in Ex. 16, find the corresponding inequalities delimiting the m-ary products of latent roots of A, with respect to those of $\bar{A}'A$, $m = 2, 3, \ldots, n$.

18. Each latent root λ of any principal minor of a unitary matrix is such that $|\lambda| \leqslant 1$. (Loewy, Brauer.)

[Partition the identity $\bar{A}'A = I$, characterizing a unitary matrix A, as follows:

$$\begin{bmatrix} \bar{A}'_{mm} & \bar{A}'_{pm} \\ \bar{A}'_{mp} & \bar{A}'_{pp} \end{bmatrix} \begin{bmatrix} A_{mm} & A_{mp} \\ A_{pm} & A_{pp} \end{bmatrix} = \begin{bmatrix} I_{mm} & \cdot \\ \cdot & I_{pp} \end{bmatrix}, \quad p = n - m,$$

where suffixes represent orders of submatrices. The leading element of the product on the left is

$$\bar{A}'_{mm}A_{mm} + \bar{A}'_{pm}A_{pm} = I_{mm}.$$

All the matrices (cf. Ex. 16) in this identity are non-negative definite. Hence, if z is a latent point of A_{mm}, then

$$\bar{z}'\bar{A}'_{mm}A_{mm}z \leqslant \bar{z}'I_{mm}z, \quad \text{i.e.} \leqslant \bar{z}'z.$$

But (cf. Ex. 3) $$\bar{z}'A'_{mm}\bar{A}_{mm}z = \bar{\lambda}\lambda\bar{z}'z.$$

Hence $$\bar{\lambda}\lambda \leqslant 1.]$$

19. If $\bar{A} = A^{-1}$, the latent roots of A are of the form $e^{i\alpha}$.

[Take $Az = \lambda z$, $uA = \lambda u$, $z \neq 0$, $u \neq 0$. Then $\bar{u}\bar{A}Az = \bar{u}z = \bar{\lambda}\lambda\bar{u}z$.]

4. **Historical Note.**—The reality of the latent roots of a real symmetric matrix was first established by Cauchy in 1829; *Œuvres*, IX (2), 172–5. Sylvester in 1853 considered the general secular equation; *Coll. Works*, I, 634. The extension to the Hermitian case was noted by Hermite, *C. R.*, **41** (1855), 181–3, or *Œuvres*, I, 479–81. Proofs are found in Clebsch, *J. für Math.*, **57** (1860), 319, and **62** (1863), 328–9; also in Christoffel, *J. für Math.*, **63** (1864), 255, where the linear nature of the invariant factors is first proved.

The theorem that the latent roots of an orthogonal matrix have unit modulus is due to Brioschi; *J. de Math.*, **19** (1854), 253–6; or *Opere Mat.*, V, 161–4.

The unitary reduction of a general complex matrix to the form Γ of Jacobi was given by Schur, *Math. Ann.*, **66** (1909), 488–510.

The theorem of Ex. 15, on delimiting the latent roots of a matrix, was given in part by Bendixson in 1900; *Öfversigt. Vetenskaps-Akad. Forh.* (Stockholm), 1900, No. 9, 1099–1103, and *Acta Math.*, **25** (1902), 359–65. Bendixson's result, which concerned the real part of a latent root of a real matrix, was extended by A. Hirsch to complex matrices; *ibid.*, 367–70. Bromwich gave the result for real and imaginary parts of latent roots in the general case; *Report Brit. Assoc.*, **74** (1904), 440–1, and *Acta Math.*, **33** (1906), 297–304. The theorem of Ex. 16 on the latent roots of A and $\bar{A}'A$ is due to E. T. Browne; *Bull. Amer. Math. Soc.*, **34** (1928), 363; see also **36** (1930), 205–10. Further papers by E. T. Browne, *Amer. Math. Monthly*, **46** (1939), 252–65, and W. V. Parker, *Duke Math. Journal*, **3** (1937), 484–7, and **10** (1943), 479–82, may be consulted.

The progress of the theory of orthogonants, i.e. the determinants of orthogonal matrices, may be followed in Muir's historical volumes, completed by a memoir bringing the commentary up to 1920, *Proc. Roy. Soc. Edin.*, **47** (1926), 252–82. Of special relevance is the account of Cayley's orthogonant, *J. für Math.*, **32** (1846), 119–23, or *Coll. Works*, I, 332–6, and of the various other attempts to construct a general orthogonal matrix.

CHAPTER IX

THE CANONICAL REDUCTION OF PENCILS OF MATRICES

The characteristic matrix, or rather family of matrices, $A - \lambda I$, which plays so important a part in the collineatory reduction of a matrix A, is a particular case of a *matrix pencil*. If we have a set of matrices A_i, each of which is equivalent in a certain field $\mathcal{F}$ to one of another set of matrices B_i, all of the same order, the relation of equivalence being the same for all, namely,

$$PA_iQ = B_i, \quad |P| \neq 0, \quad |Q| \neq 0, \quad i = 1, 2, 3, \ldots, \qquad (1)$$

then any linear combination of the A_i with scalar parameters λ_i is equivalent to the same combination of the B_i, so that

$$\lambda_1 B_1 + \lambda_2 B_2 + \ldots + \lambda_\nu B_\nu = P(\lambda_1 A_1 + \lambda_2 A_2 + \ldots + \lambda_\nu A_\nu)Q. \quad (2)$$

The matrices $\Sigma\lambda_i A_i$ and $\Sigma\lambda_i B_i$ will be termed *combinantal matrices* or *combinants*, equivalent in $\mathcal{F}$. A *basis* for such a combinant is a set of ν matrices A_i linearly independent in $\mathcal{F}$, by which it is implied that no relation $\sum_{i=1}^{\nu} a_i A_i = 0$ exists, where the a_i are scalar numbers in $\mathcal{F}$, not all zero. A combinant resembles a λ-matrix, but its elements are linear and homogeneous in ν variables. If $\nu = 2$ the combinant $\lambda A_1 + \mu A_2$ is called a matrix *pencil*, and is based on a pair of matrices A_1, A_2, neither of which is null or a scalar multiple of the other. The scalar parameters λ, μ are arbitrary; if we fix them in any particular instance the matrix so obtained is termed a member of the pencil. If $\nu = 3$ the system is called a *net*. Our present concern is with equivalent pencils of matrices, and the associated pencils of bilinear and quadratic forms.

In Chapter III the elements of the matrices P and Q were functions of λ: here this is no longer the case. The elements of P and Q, like those of the A_i and B_i, are constants in $\mathcal{F}$, and are therefore *independent* of the parameters λ, μ. If each A_i, B_i has n' rows and n columns, P and Q are square matrices of respective orders n' and n.

EXAMPLES

1. The pencil $\lambda A + \mu I$ is equivalent to the pencil $\lambda B + \mu I$ if, and only if, $PQ = I$, that is, only in the collineatory group.

[Thus the theory of the collineatory transformation of a single matrix, the topic of Chapters V and VI, is contained in the theory of the equivalent transformation of this special type of pencil.]

2. The pencils of bilinear forms $u(\lambda A_1 + \mu A_2)x$ and $v(\lambda B_1 + \mu B_2)y$ are equivalent if $v = uP$, $y = Qx$.

3. If the variables u, x are contragredient, pencils of collineations are equi valent, provided that $PQ = I$.

4. Pencils of Hermitian and quadratic forms are equivalent, provided that $\tilde{P}' = Q$.

1. Singular and Non-Singular Pencils.

Within the field of all complex numbers every combinantal system contains singular members. In particular a pencil of square matrices $\Lambda = \lambda A_1 + \mu A_2$ of order $n \times n$ contains singular members, for which

$$\Delta = |\Lambda| = |\lambda A_1 + \mu A_2| = 0. \quad . \quad . \quad . \quad . \quad (3)$$

Each m-rowed minor of this determinant Δ is evidently a binary m-ic in λ, μ, while the determinant Δ itself is a binary n-ic,

$$\Delta = \Delta(\lambda, \mu) = p_0\lambda^n + p_1\lambda^{n-1}\mu + \ldots + p_n\mu^n, \quad . \quad (4)$$

where, in particular, $p_0 = |A_1|$, $p_n = |A_2|$. If the determinant is resolved into homogeneous linear factors,

$$\Delta = (\alpha_1\lambda - \beta_1\mu)(\alpha_2\lambda - \beta_2\mu)\ldots(\alpha_n\lambda - \beta_n\mu), \quad . \quad (5)$$

then the singular members of the pencil are given by $\beta_i A_1 + \alpha_i A_2$. There are thus n or less really distinct singular members, one for each different value of the ratio $\lambda : \mu$ which causes Δ to vanish. All other members are non-singular.

When the coefficients p_i in (4) above do not all vanish we shall call the pencil Λ *non-singular*. The pencil is called *singular* when, and only when, the determinant Δ vanishes identically for all values of λ, μ; so that each $p_i = 0$, and every member of the pencil is singular. Such is the case for square matrices: other rectangular matrices and combinantal systems are always singular.

EXAMPLES

1. The pencil of conics $\lambda x^2 + (\lambda + \mu)y^2 + \mu z^2 = 0$ is non-singular, since

$$\Delta = \begin{vmatrix} \lambda & \cdot & \cdot \\ \cdot & \lambda + \mu & \cdot \\ \cdot & \cdot & \mu \end{vmatrix} = \lambda\mu(\lambda + \mu) \neq 0.$$

It possesses, however, three singular members $x^2 + y^2 = 0$, $y^2 + z^2 = 0$, and $x^2 - z^2 = 0$, corresponding to $\mu = 0$, $\lambda = 0$, $\lambda + \mu = 0$.

2. The pencil of conics $\lambda xz + \mu yz = 0$ is singular, since

$$\Delta = \begin{vmatrix} \cdot & \cdot & \lambda \\ \cdot & \cdot & \mu \\ \lambda & \mu & \cdot \end{vmatrix} = 0.$$

[Every member of the pencil is singular: the pencil consists of a fixed straight line and a variable straight line through a fixed point.]

2. Equivalent Canonical Reduction in the Non-Singular Case.

A non-singular pencil of matrices can be reduced without difficulty by equivalent transformations to rational or irrational canonical forms, which depend on the corresponding collineatory forms for a single matrix. If the pencil is $\Lambda = \lambda A_1 + \mu A_2$, then one or other, or both, of the matrices A_1 and A_2 may be singular; it is always possible, however, if the pencil itself is non-singular, to change the basis to a pair of matrices D_1, D_2, neither of which is singular. For if it be ascertained what values of the ratio $\lambda : \mu$ yield singular members of the pencil Λ, we may then choose (in $\mathcal{F}$) two other distinct ratios $\lambda_1 : \mu_1$ and $\lambda_2 : \mu_2$ differing from these, and may take as new basis

$$\left.\begin{aligned} &D_1 = \lambda_1 A_1 + \mu_1 A_2, \quad D_2 = \lambda_2 A_1 + \mu_2 A_2, \\ &|D_1| \neq 0, \quad |D_2| \neq 0, \quad \lambda_1\mu_2 - \lambda_2\mu_1 \neq 0. \end{aligned}\right\} \quad . \quad . \quad (6)$$

In terms of the new basis we shall have

$$\lambda A_1 + \mu A_2 = \rho D_1 + \sigma D_2,$$

where

$$\lambda = \lambda_1\rho + \lambda_2\sigma, \quad \mu = \mu_1\rho + \mu_2\sigma, \quad . \quad . \quad . \quad (7)$$

or

$$\begin{bmatrix} \lambda \\ \mu \end{bmatrix} = \begin{bmatrix} \lambda_1 & \lambda_2 \\ \mu_1 & \mu_2 \end{bmatrix} \begin{bmatrix} \rho \\ \sigma \end{bmatrix}.$$

By taking $P = I$, $Q = D_2^{-1}$, which is possible, since $|D_2| \neq 0$ we transform $\rho D_1 + \sigma D_2$ to $\rho D + \sigma I$, where $D = D_1 D_2^{-1}$. The matrix D can be further reduced by collineatory transformations, either to the rational form B of Chapter V (p. 49), or to the classical form C of Chapter VI (p. 61), whichever is desired. For example if $HDH^{-1} = C$, we now have

$$H(\rho D_1 + \sigma D_2)D_2^{-1}H^{-1} = \rho C + \sigma I, \quad . \quad . \quad . \quad (8)$$

so that the pencil Λ has been reduced to a simple canonical form, from which the canonical shape of any member can be ascertained by substituting the appropriate values of ρ, σ.

It is of course to be observed that at this stage the reduction is applicable, by the insertion of variables u and x, to bilinear forms, but not to quadratic or Hermitian forms, since the matrices H and $D_2^{-1}H^{-1}$ are not in general the transposed or associates of each other.

EXAMPLES

1. The typical isolated submatrix in the rational canonical form of a non singular pencil has an appearance such as, e.g., $p = 4$,

$$B_p(\lambda, \mu) \equiv \lambda B + \mu I = \begin{bmatrix} \mu & \lambda & \cdot & \cdot \\ \cdot & \mu & \lambda & \cdot \\ \cdot & \cdot & \mu & \lambda \\ b_4\lambda & b_3\lambda & b_2\lambda & \mu + b_1\lambda \end{bmatrix}.$$

2. An alternative form exists in which the rôles of λ and μ are interchanged, the coefficients b_i being suitably modified; and two further forms may be obtained by transposition.

3. The typical submatrix in the classical form of a non-singular pencil is, e.g.,

$$C_p(\lambda, \mu) \equiv \lambda C + \mu I = \begin{bmatrix} a\lambda + \mu & \lambda & \cdot & \cdot \\ \cdot & a\lambda + \mu & \lambda & \cdot \\ \cdot & \cdot & a\lambda + \mu & \lambda \\ \cdot & \cdot & \cdot & a\lambda + \mu \end{bmatrix}.$$

3. The Invariant Factors of a Matrix Pencil.

In many respects the pencil $\Lambda = \lambda A_1 + \mu A_2$ is related to the characteristic family $A - \lambda I$ of earlier chapters in much the same way as a binary form is related to a polynomial in a single variable; the properties are analogous, and any additional complexity of statement is largely on the surface. It may be surmised therefore that a pencil possesses binary invariant factors and elementary divisors which, like those of the simpler family $A - \lambda I$, are unaltered under equivalent transformations. This is actually the case.

For example, by rendering the work on p. 16 homogeneous we have, let us say

$$L = [\lambda^3 + 2\lambda^2\mu + 2\lambda\mu^2 + \mu^3,\ 3\lambda^3 + 8\lambda^2\mu + 9\lambda\mu^2 + 4\mu^3],$$
$$LQ = [\lambda + \mu,\ \cdot],$$

where Q is a second-order square matrix whose elements are homogeneous binary forms (in this case, quadratics) in λ, μ.

Here, in effect, one has replaced the original λ by λ/μ, and has cleared a common denominator—a power of μ—from all the terms. It is apparent that all essential properties of invariant factors (pp. 23–7) are preserved: so that, by Smith's method of reduction for λ-matrices (p. 23), a set of invariant factors

$$\Delta_0/\Delta_1, \quad \Delta_1/\Delta_2, \quad \ldots$$

can be constructed, where Δ_0, Δ_1, Δ_2, ... are binary forms in λ, μ, with coefficients in $\mathcal{F}$, such that each Δ_i contains its successor as factor. Each Δ_i can, as before, also be determined as the H.C.F. of the minors of a certain order k, in the original array Λ: and this applies whether Λ is singular or not. This k takes the successive values ρ, $\rho - 1$, $\rho - 2$, ... , where ρ is the rank of Λ. For the non-singular case, when Λ has orders $n \times n$, the determinant $\Delta = |\Lambda|$ does not vanish identically in λ, μ, so that $\rho = n$, and $\Delta_0 = \Delta$ itself.

If we therefore adapt the formula (12), p. 23, we may write

$$P\Lambda Q = \text{diag}\,(E_1(\lambda, \mu), \quad E_2(\lambda, \mu), \ldots, \quad E_\rho(\lambda, \mu), \cdot, \ldots \cdot),$$

where it is important again to notice that in general P and Q are not constant matrices but involve λ, μ. On the other hand, by restricting P and Q to be constants we shall render the canonical form $P\Lambda Q$ more complicated than this diagonal matrix: but inasmuch as the restricted $P\Lambda Q$ is equivalent to Λ its invariant factors will again be identical with the products (p. 27)

$$G_k(\lambda, \mu) = E_1(\lambda, \mu)\, E_2(\lambda, \mu) \ldots E_k(\lambda, \mu).$$

4. Invariance under Change of Basis.

The invariant factors are not merely invariant under equivalent transformations of the pencil in the field: they are invariant (or rather *covariant*) for the change of basis induced by a non-singular linear transformation (7) from λ, μ to ρ, σ. This is due to the elementary fact that such a non-singular binary transformation can be carried out in each partial quotient and remainder in the H.C.F. process for two binary polynomials ϕ and ψ without affecting properties of divisibility, or the degree of any remainder, and without making any intermediate remainder vanish: the transformation can in fact be made at any stage whatever of the process without altering the final result. Thus the invariant factors in terms of ρ and σ of the pencil Λ referred to a new basis are simply obtained by making the substitution from λ, μ to ρ, σ in the invariant factors of the original pencil.

In fact if $\theta = hg$, $\phi = kg$ denote two minors whose H.C.F. g is required, then θ, ϕ are homogeneous and of the same degree, let us say p, in λ, μ, while g is homogeneous and of degree q. On transforming λ, μ to ρ, σ, this g is manifestly a common factor of degree p in ρ, σ. Hence if g' is the H.C.F. of θ and ϕ, quâ functions of ρ, σ, then g' contains g. But conversely, on interchanging the rôles of λ, μ and ρ, σ, the H.C.F. g contains g'. Hence they are, to a constant numerical factor, identical polynomials.

By resolving the invariant factors into homogeneous linear factors, in general by an irrational process, we arrive at the concept of the *elementary divisors* of a pencil, which, being simply the binary homogeneous analogue of the elementary divisors as earlier defined, hardly call for further explanation. Pencils of matrices are equivalent in a field $\mathcal{F}$ if, and only if, they have the same invariant factors or, alternatively, the same elementary divisors.

EXAMPLES

1. Two pencils of bilinear forms $\lambda u A_1 x + \mu u A_2 x$ and $\lambda v B_1 y + \mu v B_2 y$ are equivalent in $\mathcal{F}$ if, and only if, the matrix pencils $\lambda A_1 + \mu A_2$ and $\lambda B_1 + \mu B_2$ have the same invariant factors.

2. Write out in full the canonical forms, rational and classical, of a pencil of bilinear forms, non-singular, elementary and of the fourth order.

3. The invariant factors of the pair of conics

$$x'Ax = x^2 + y^2 \quad \text{and} \quad x'Bx = y^2 + z^2,$$

for which the matrix pencil is

$$\lambda A + \mu B = \begin{bmatrix} \lambda & & \\ & \lambda+\mu & \\ & & \mu \end{bmatrix},$$

are $\lambda\mu(\lambda+\mu)$, 1, 1. Two other conics of the system are $x^2 + 2y^2 + z^2$ and $x^2 - z^2$ with invariant factors $2\rho(\rho^2 - \sigma^2)$, 1, 1, where $\lambda = \rho + \sigma$, $\mu = \rho - \sigma$.

4. Given that the invariant factors of a certain pencil of bilinear quinary forms are $\lambda^3 + \mu^3$, $\lambda^2 - \lambda\mu + \mu^2$, find the canonical form of the pencil.

[The corresponding submatrices in the rational canonical form $B(\lambda, \mu)$ must be

$$\begin{bmatrix} \lambda & \mu & \cdot \\ \cdot & \lambda & \mu \\ \mu & \cdot & \lambda \end{bmatrix} \quad \text{and} \quad \begin{bmatrix} \lambda & \mu \\ -\mu & \lambda-\mu \end{bmatrix},$$

so that the most general pencil in question is $x'P(\lambda I + \mu B)Qy$, where $|P| \neq 0$, $|Q| \neq 0$, and where

$$B = \begin{bmatrix} \cdot & 1 & \cdot & \cdot & \cdot \\ \cdot & \cdot & 1 & \cdot & \cdot \\ 1 & \cdot & \cdot & \cdot & \cdot \\ \cdot & \cdot & \cdot & \cdot & 1 \\ \cdot & \cdot & \cdot & -1 & -1 \end{bmatrix}.$$

The canonical form of the pencil is therefore

$$\mu(\xi_1\eta_2 + \xi_2\eta_3 + \xi_3\eta_1 + \xi_4\eta_5 - \xi_5\eta_4 - \xi_5\eta_5) \\ + \lambda(\xi_1\eta_1 + \xi_2\eta_2 + \xi_3\eta_3 + \xi_4\eta_4 + \xi_5\eta_5).]$$

5. The singular members of the above pencil are given by values of λ, μ which cause the highest invariant factor to vanish.

6. A rational singular member of the pencil in question is

$$\xi_1(\eta_1 - \eta_2) + \xi_2(\eta_2 - \eta_3) + \xi_3(\eta_3 - \eta_1) + \xi_4(\eta_4 - \eta_5) + \xi_5(\eta_4 + 2\eta_5).$$

7. *The invariant factors of a pencil of Hermitian matrices are real.*

[The H.C.F. of minors of order m in the determinant Δ of the pencil is a homogeneous polynomial in λ, μ. By transposing the pencil we find that the H.C.F. is equally well the complex conjugate of the above. But transposition leaves unaltered the value of each minor and therefore the value of the H.C.F. The H.C.F., being its own complex conjugate, must therefore be real: thus the invariant factors are also real.]

5. The Dependence of Vectors with Binary Linear Elements. Minimal Indices.

The foregoing method in § 2, p. 115, of reducing pencils of matrices to equivalent canonical form, applies only to non-singular pencils. Where the matrix pencil is square but singular, or else is rectangular, peculiar features arise, necessitating a closer examination of the dependence in $\mathcal{F}$ of the row and column vectors which comprise the matrix.

We start with the fact that when a pencil (or combinant) Λ is singular there exist vectors u, x in $\mathcal{F}$ such that at least one and perhaps both of the conditions

$$u\Lambda = 0,\ u \neq 0;\quad \Lambda x = 0,\ x \neq 0, \qquad . \quad . \quad . \quad (9)$$

is satisfied. If Λ has n' rows and n columns, these conditions may also be written

$$\sum_{i=1}^{n'} u_i \text{ row}_i = 0,\quad \sum_{j=1}^{n} x_j \text{ col}_j = 0. \quad . \quad . \quad . \quad . \quad (10)$$

If $n' > n$, then $u\Lambda = 0$ is certainly satisfied (Ex. 4, p. 46): if $n' < n$, $\Lambda x = 0$ is certainly satisfied. If $n' = n$, then $|\Lambda| = 0$, and both are satisfied. These statements are coextensive with the definition of a singular pencil Λ on p. 114. We must now consider the character of these dependence-vectors u and x. Consider, for example, the following three binary vectors

$$p = [\lambda + \mu,\ \lambda],\quad q = [\mu,\ \lambda + \mu],\quad r = [\lambda + \mu,\ \lambda + \mu].$$

It may be verified that

$$(\lambda^2 + \lambda\mu)p + (\lambda\mu + \mu^2)q - (\lambda^2 + \lambda\mu + \mu^2)r = [0, 0],$$

and that no relation of dependence holds for p, q, r with coefficients of less than the second degree in λ, μ. The vectors p, q, r will be said to possess *dependence of the second order* in λ, μ. Alternatively we may weld the vectors into a rectangular matrix Λ, and write the relation of dependence in the shape

$$[\lambda^2 + \lambda\mu,\ \lambda\mu + \mu^2,\ -\lambda^2 - \lambda\mu - \mu^2]\begin{bmatrix}\lambda+\mu & \lambda \\ \mu & \lambda+\mu \\ \lambda+\mu & \lambda+\mu\end{bmatrix} = [0, 0], \quad (11)$$

or, let us say, $u\Lambda = 0$. In this example the columns of the matrix Λ are independent. On the other hand—to take another example—in the matrix

$$\Lambda = \begin{bmatrix}\lambda & \lambda \\ \mu & \mu\end{bmatrix},$$

the columns have dependence of zero order, expressed by $\Lambda x = 0$, where $x = \{1, -1\}$, while the rows have dependence of the first order, since $[\mu, -\lambda]\Lambda = 0$. Evidently dependence of order m implies dependence of every higher order $m + k$, since in the relations of dependence $u\Lambda = 0$ (or $\Lambda x = 0$) we may multiply u (or x) throughout by any homogeneous scalar polynomial in λ, μ and obtain similar relations. Also, since a non-homogeneous identity in λ, μ implies that each of its distinct homogeneous parts is an identity, there is no loss of generality in confining the discussion to the homogeneous identity of any order. But it is the *minimal* order of dependence that is important, for the reason that it possesses invariant properties.

Indeed *the minimal order* or *index* (which we shall denote by m in the case of row-dependence expressed by $u\Lambda = 0$, and by m' in the case of column dependence expressed by $\Lambda x = 0$) *is invariant under two kinds of transformation,*

(i) equivalent transformation of the matrix pencil Λ,

(ii) homogeneous non-singular linear transformation of λ, μ to ρ, σ; in a word, change of basis.

In proof of (i), consider an equivalent matrix $P\Lambda Q$, $|P| \neq 0$, $|Q| \neq 0$, and suppose it possible that $vP\Lambda Q = 0$, where v is a vector of minimal index, less than m, annulling $P\Lambda Q$. Since $|Q| \neq 0$, we must have $vP\Lambda QQ^{-1} = vP\Lambda = 0$, where, since the elements of P are constants, the vector vP annulling Λ must be, like v, of lower index than m. But this contradicts the assumption that u is of minimal degree. Hence v cannot be of lower index than u: and by the reflexive nature of equivalent transformations it follows equally that u cannot be of lower index than v. Thus the minimal index m of u is invariant under equivalent transformations of Λ; and (i) is established.

In proof of (ii), we note that linear transformation of λ, μ to ρ, σ *cannot conceivably raise* the degree in the transformed vector u, though it might lead to the lowering of it, through the cancelling of some common factor in the transformed elements; in such a case, however, retransformation from ρ, σ to λ, μ, which is possible since the first transformation was non-singular, would lead to a vector of lower degree than u, which is again contrary to hypothesis. Hence the minimal index m is invariant under change of basis, and (ii) is established. By the same reasoning applied to the transposed pencil the order m' is likewise invariant under (i) and (ii).

It is to be observed that, in all the above, the matrix pencil Λ is not necessarily square, but may be rectangular, in which case P and Q are square but of different orders.

6. The Canonical Minimal Submatrix, and the Vector of Apolarity.

By equivalent transformations we shall reduce a given minimal relation $u\Lambda = 0$ to $uP^{-1}P\Lambda Q = 0$, thereby bringing the vector u to the form uP^{-1}, and the pencil Λ to the form $P\Lambda Q$. A lemma can now be proved which introduces a canonical vector

$$\omega_{(m)} = [\mu^m, -\lambda\mu^{m-1}, \lambda^2\mu^{m-2}, \ldots, (-)^m \lambda^m, 0, \ldots, 0], \quad m \geqslant 0, \quad (12)$$

called *the vector of apolarity* and consisting of $m + 1$ non-zero terms arranged in descending powers of μ, followed by zeros. There will also occur a canonical singular submatrix,

$$L_m = \begin{bmatrix} \lambda & & & & \\ \mu & \lambda & & & \\ & \mu & \ddots & & \\ & & \ddots & \ddots & \\ & & & \ddots & \lambda \\ & & & & \mu \end{bmatrix}, \quad \ldots \ldots \quad (13)$$

of $m+1$ rows and m columns, containing twin diagonals each of m equal elements λ or μ. In particular

$$\omega_{(0)} = [1, 0, \ldots, 0], \quad L_1 = \begin{bmatrix} \lambda \\ \mu \end{bmatrix}, \quad L_2 = \begin{bmatrix} \lambda & \cdot \\ \mu & \lambda \\ \cdot & \mu \end{bmatrix}, \ldots \quad (14)$$

Lemma I.—*There exist in* $\mathcal{F}$ *non-singular constant matrices* P *and* Q *which reduce a given minimal relation* $u\Lambda = 0$ *to the form*

$$uP^{-1}P\Lambda Q \equiv \omega_{(m)} \begin{bmatrix} L_m & \cdot \\ \mu R & \Lambda_0 \end{bmatrix} = 0, \quad . \quad . \quad . \quad (15)$$

where Λ_0 *is a pencil with* $m+1$ *fewer rows and* m *fewer columns than* Λ, *and where* R *is a submatrix with constant elements. If* $m = 0$ *the* L_m *and* μR *are non-existent (as in* (19) *below).*

Proof.—We build suitable matrices P and Q, step by step. We first note that at least one of the elements of u must contain a term in μ^m, otherwise we could divide through by λ and obtain a vector of lower index annulling Λ. By an interchange of elements, that is to say, of *columns*, of u, we bring this term into the first element, dividing it if need be so as to give μ^m unity for coefficient. Next, by applying to u the operations $\text{col}_j - k\,\text{col}_1$, $(j > 1)$, we may use this μ^m to cast out any other μ^m from the remaining elements of u. These operations are equivalent to a post-multiplication by a certain matrix H, which we shall also call P_1^{-1}: and we can now write

$$\begin{aligned} u_{(1)} &= uP_1^{-1} \\ &= [\mu^m + a\mu^{m-1}\lambda + \ldots, \ \beta\mu^{m-1}\lambda + \ldots, \ \gamma\mu^{m-1}\lambda + \ldots, \ \ldots] \\ &= [\mu^m \bmod \lambda, \ \ \lambda v], \quad . \quad . \quad . \quad . \quad . \quad . \quad . \quad (16) \end{aligned}$$

where λv denotes the vector deprived of its first element, so that v itself is a vector whose $(n'-1)$ elements are homogeneous polynomials, in λ and μ, of order $m-1$. If a_i denotes the original coefficient of μ^m in the ith element, then evidently

$$[a_1, a_2, \ldots, a_i, \ldots]\, H = [1, 0, \ldots, 0], \quad . \quad . \quad (17)$$

and we shall say that the process has *concentrated* the vector a in leading element.

We have thus reduced the original identity $u\Lambda = 0$ to the form

$$0 = uP_1^{-1}P_1\Lambda = u_{(1)}\Lambda_1, \quad . \quad . \quad . \quad . \quad . \quad (18)$$

involving a vector $u_{(1)} = uP_1^{-1}$, and a pencil $\Lambda_1 = P_1\Lambda$.

In the top row of the new pencil Λ_1 no term in μ can occur, for there is no term in the identity (18) to cancel the μ^{m+1} which would then arise. Hence the top row is either null or else consists of terms in λ only. If it is null we have at once $[1, 0, \ldots, 0]\Lambda_1 = 0$; and this can only happen if $m = 0$. This proves the initial stage of the Lemma, yielding a reduction

$$m = 0, \quad u\Lambda = \omega_{(0)}\begin{bmatrix} \cdot \\ \Lambda_0 \end{bmatrix} = 0. \quad . \quad . \quad . \quad (19)$$

If, however, $m > 0$, the non-zero elements in the first row of Λ_1 may be concentrated in the leading element λ by means of a post-multiplier Q_1, just as in (17) above. Thus

$$\Lambda_1 Q_1 = \begin{bmatrix} \lambda, 0, \ldots & 0 \\ \cdot \quad \cdot \quad \cdot \quad \cdot & \cdot \end{bmatrix} = \Lambda_2, \quad . \quad . \quad . \quad . \quad (20)$$

and we have secured the first row of the desired form L_m. A set of premultiplications P_2, involving operations

$$\text{row}_i - a_i \,\text{row}_1, \qquad (i > 1),$$

may now be invoked to concentrate in the leading element λ all terms in λ which occur in the first column of Λ_2, so that the rest of the column is now free from λ. The reciprocal postmultiplications P_2^{-1}, namely

$$\text{col}_1 + a_i \,\text{col}_i, \qquad (i > 1)$$

will then affect the vector $u_{(1)}$. But since no term in μ^m exists except at the first element, this reciprocal process leaves the character of the vector in (16) unaltered. We can therefore write

$$0 = u_{(1)} P_2^{-1} P_2 \Lambda_2 = u_{(2)} \Lambda_3, \quad . \quad . \quad . \quad . \quad (21)$$

where $\quad u_{(1)} P_2^{-1} = u_{(2)} = [\mu^m \bmod \lambda,\ \lambda v\rfloor, \quad \Lambda_3 = \begin{bmatrix} \lambda & \cdot \\ \mu c & \Gamma \end{bmatrix}.$

Here μc denotes the $(n' - 1)$ further elements of the first column, below the concentrated λ, and Γ denotes a submatrix pencil of $(n' - 1)$ rows and $(n - 1)$ columns.

If $v = 0$, then on dividing the vector $u_{(2)}$ throughout by its first (and non-zero) element we should have $[1, 0, \ldots, 0]\Lambda_3 = 0$, which is impossible since $m > 0$. Hence $v \neq 0$. On forming the product $u_{(2)}\Lambda_3$ in (21) it follows that

$$v\Gamma = 0, \quad v \neq 0. \quad . \quad . \quad . \quad . \quad . \quad . \quad . \quad (22)$$

Now v is a minimal vector of index $m-1$ annulling Γ. For if not, a minimal w of index $p(< m-1)$ must exist. Construct the vector $[\mu wc, -\lambda w]$, where we note that the first term is scalar. This non-zero vector is of index $p+1(< m)$ and annuls Λ_3, as expansion at once shows: which is impossible. Hence v is a true minimal.

Assuming the truth of the Lemma (already demonstrated when $m=0$) for the index $p=m-1$, and for each pencil of n' or fewer rows, we conclude the proof by induction. For assuming that sub-matrices P_0 and Q_0 exist which reduce the minimal relation $v\Gamma=0$ to the form (15), with $\omega_{(p)}$, L_p replacing $\omega_{(m)}$, L_m, let the corresponding matrices

$$P^{-1}=\begin{bmatrix}1 & \cdot \\ \cdot & -P_0^{-1}\end{bmatrix},\quad P=\begin{bmatrix}1 & \cdot \\ \cdot & -P_0\end{bmatrix},\quad Q=\begin{bmatrix}1 & \cdot \\ \cdot & -Q_0\end{bmatrix}$$

act upon the identity (21). Then

$$0=u_{(2)}P^{-1}\,.\,P\Lambda_3Q=[u_1,\ -\lambda\omega_{(p)}]\begin{bmatrix}\lambda & \cdot \\ \mu d & P_0\Gamma Q_0\end{bmatrix}, \qquad (23)$$

where $d=-P_0c$ is the new form of the set of coefficients of μ in the first column, and where $vP_0^{-1}=\omega_{(p)}$, and u_1 is the leading component of $u_{(2)}$. The first $p+1$ rows of the pencil can now be written

$$\begin{bmatrix}\lambda & \cdot & \cdot & \cdots & & \cdot \\ \mu\delta_2 & \lambda & & & & \\ \mu\delta_3 & \mu & \ddots & & & \\ \vdots & & \ddots & \ddots & & \\ \vdots & & & \ddots & \lambda & \\ \mu\delta_{p+1} & & & \cdot & \mu & \cdot\ldots\cdot\end{bmatrix}, \quad \ldots\ldots \quad (24)$$

while the later rows have assumed the desired form $[\mu R, \Lambda_0]$ of (15).

Without otherwise disturbing the pencil, we can remove each δ_{i+1}, except δ_2, by the operation

$$\text{col}_1-\delta_{i+1}\,\text{col}_i,\quad \text{row}_i+\delta_{i+1}\,\text{row}_1,\qquad (i>1)\quad .\quad (25)$$

which induces the reciprocal $\text{col}_1-\delta_{i+1}\,\text{col}_i$, $(i>1)$ upon the minimal vector, and evidently only changes its leading element, let us say, to u_1', but again leaves the concentrated μ^m unchanged. The identity (23), so far as it concerns the first column of the final matrix, now reads

$$[u_1',\ -\lambda\omega_{(p)}]\{\lambda,\ \mu\delta_2,\ 0,\ \ldots,\ 0\}=0,\qquad .\ .\ (26)$$

which is only satisfied if $u_1' = \mu^{p+1}\delta_2$, since the leading term of $\omega_{(p)}$ is μ^p. But $p = m - 1$. Hence $\delta_2 = 1$, $u_1' = \mu^m$. The operation (25) has in fact stripped u_1 of all but its leading element, and (26) becomes $\omega_{(m)}\{\lambda, \mu, 0, \ldots, 0\} = 0$. At the same time the submatrix (24) becomes $[L_m \cdot]$, the relation (15) being thus secured. This completes the proof of the Lemma.

Lemma II.—*If* $\Lambda x = 0$, *the vector* x *being of minimal index* m′, *then the pencil* Λ *can be reduced to a form in which the first* m′ + 1 *columns are* $L_{m'}$ *and zeros, where* L′ *is the transposed of* L.

This follows by applying the argument of Lemma I to columns instead of rows.

The submatrix L_m of (13), which we may pause to examine, is typical of the canonical form of a singular pencil. It differs from previous canonical submatrices in being rectangular, $m + 1 \times m$, instead of square. Further, though its rows possess dependence of minimal index m, *its columns are independent* in λ, μ, as can readily be verified: they would be dependent were it not for the semi-isolated λ in the first row and the μ in the last row.

The application of these Lemmas to the rational reduction of a singular pencil will now be considered.

7. The Rational Reduction of a Singular Pencil.

Theorem I.—*The matrix* $\Lambda = [\lambda a_{ij} + \mu b_{ij}]$ *of a singular pencil can be reduced by rational equivalent transformations to the canonical form*

$$P\Lambda Q = L = \begin{bmatrix} L_{m_1} & & & & & & & \\ & L_{m_2} & & & & & & \\ & & \ddots & & & & & \\ & & & L_{m_k} & & & & \\ & & & & L'_{m_1'} & & & \\ & & & & & L'_{m_2'} & & \\ & & & & & & \ddots & \\ & & & & & & & M \end{bmatrix}, \quad . \quad (27)$$

where M *is a square non-singular core in rational canonical form* $[B_p(\lambda, \mu)]$ *or possibly, with a change of basis,* $[B_p(\rho, \sigma)]$.

Proof.—In the first place, row-dependence of zero index may be disposed of, resulting in a certain number of null rows in the canonical form. In the same way column-dependence of zero index will lead to a certain number of null columns. For convenience we shall suppose this dealt with, and shall therefore take Λ to be the remaining rows and columns. By Lemma I, if the rows have dependence of index m_1, we may reduce Λ to the shape

$$P_1 \Lambda Q_1 = \begin{bmatrix} L_{m_1} & \cdot \\ \mu R & \Lambda_1 \end{bmatrix},$$

where any row-dependence that may exist in μR is necessarily of zero order, which has been disposed of for Λ, so that any further row-dependence in these lower rows $[\mu R, \Lambda_1]$ must be sought for exclusively in the submatrix Λ_1. Moreover, such dependence must be of order $m_2 \geqslant m_1$, because of the minimal nature of m_1. Hence we proceed to reduce Λ_1 exactly as we did Λ, obtaining at this stage, as on many former occasions in earlier chapters,

$$P_2 P_1 \Lambda Q_1 Q_2 = \begin{bmatrix} L_{m_1} & \cdot & \cdot \\ * & L_{m_2} & \cdot \\ * & * & \Lambda_2 \end{bmatrix},$$

where the asterisks denote submatrices not in general null, involving μ but never λ. If further row-dependence exists in Λ_2 we proceed in the same manner, and so on, until we either exhaust the rows, or arrive at a submatrix Λ_k in which the rows are independent.

The dependence of columns is next considered. Since, as was observed earlier, the columns of $L_{m_1}, L_{m_2}, \ldots$ are independent among themselves, dependence of columns must be sought for, if at all, in Λ_k. This we reduce in the same manner as Λ, but in transposed fashion, obtaining at last, when dependence of columns has been exhausted, an equivalent canonical form for Λ, namely

$$\left[\begin{array}{cccc|cccc|c} L_{m_1} & \cdot & \cdot & & & & & & \cdot \\ * & L_{m_2} & \cdot & & & & & & \cdot \\ * & * & L_{m_3} & & & & & & \cdot \\ \vdots & & & \ddots & & & & & \vdots \\ \hline * & * & * & \ldots & L'_{m_1'} & * & * & \ldots & * \\ * & * & * & & & L'_{m_2'} & * & & * \\ * & * & * & & & & L'_{m_3'} & & * \\ \vdots & & & & & & & \ddots & \vdots \\ \hline * & * & * & \ldots & \cdot & & \cdot & \ldots & S \end{array}\right], \quad . \quad . \quad (28)$$

where S is a possibly existent non-singular core which may be reduced, through a change of basis if need be, to the rational canonical form $[B_p(\rho, \sigma)]$.

Such a step expresses each element of the L_m (and the $L_{m'}$) in terms of ρ, σ as in (7), p. 115. But by successive row (or column) transformations these submatrices can take the same form whether $\begin{bmatrix} \rho & & \cdots \\ \sigma & \rho & \\ & \sigma & \end{bmatrix}$ or $\begin{bmatrix} \rho & \sigma & & \cdots \\ & \rho & \sigma & \end{bmatrix}$, without altering their minimal numbers.

For example, if
$$\begin{bmatrix} \lambda \\ \mu \end{bmatrix} = \begin{bmatrix} \lambda_1 & \lambda_2 \\ \mu_1 & \mu_2 \end{bmatrix} \begin{bmatrix} \rho \\ \sigma \end{bmatrix},$$
find P and Q such that
$$P\begin{bmatrix} \lambda & \cdot \\ \mu & \lambda \\ \cdot & \mu \end{bmatrix}Q = \begin{bmatrix} \rho & \cdot \\ \sigma & \rho \\ \cdot & \sigma \end{bmatrix}.$$

It remains to clear away the redundant non-zero elements in the submatrices indicated by asterisks. These submatrices are aligned with (i) two L_m's, (ii) an L_m and an $L'_{m'}$, (iii) L_m and S, (iv) two $L'_{m'}$'s, (v) $L'_{m'}$ and S. Of these only cases (i), (ii), (iii) need be considered, the others being their transposed analogues. We proceed by induction.

$$\begin{array}{ccc|cccc}
\lambda & & & & & & \\
\mu & \lambda & & & & & \\
\cdot & \mu & \lambda & & & & \\
\cdot & \cdot & \mu & & & & \\
\hline
* & * & * & \lambda & & & \\
* & * & * & \mu & \lambda & & \\
* & * & * & & \mu & \lambda & \\
* & * & * & & & \mu &
\end{array}
\qquad
\begin{array}{cc|cccc}
\lambda & & & & & \\
\mu & \lambda & & & & \\
 & \mu & & & & \\
\hline
* & * & \lambda & \mu & & \\
* & * & & \lambda & \mu & \\
* & * & & & \lambda & \mu
\end{array}
\qquad
\begin{array}{cc|ccc}
\lambda & & & & \\
\mu & \lambda & & & \\
\cdot & \mu & & & \\
\hline
* & * & \lambda & \mu & \\
* & * & \cdot & \lambda & \mu \\
* & * & b_3\mu & b_2\mu & \lambda+b_1\mu
\end{array}$$

(i) (ii) (iii)

In (i) let a diagonal barrier be drawn just below the leading diagonal of the submatrix of asterisks, which has $m_2 + 1$ rows and m_1 columns, being aligned, let us say, with L_{m_1} above and L_{m_2} to the right. Since by hypothesis $m_1 \leqslant m_2$, the top row of asterisks lies entirely above, and the bottom row entirely below, the barrier.

Each non-zero (ijth) asterisk represents a term $a\mu$. If it is situated *above* the barrier, but not in the final column of asterisks, it may be removed *one step diagonally downwards to the right*. If it lie in the final column it may at once be deleted. This is done by utilizing the μ of L_{m_1} in its own jth column and the λ of L_{m_2} in its own ith row.

For example: $\text{row}_5 - \alpha\,\text{row}_2$, $\text{col}_2 + \alpha\,\text{col}_4$ moves a term $\alpha\mu$ from position (5, 1) to (6, 2) without disturbing anything else. The process, $\text{row}_7 - \alpha\,\text{row}_4$ deletes $\alpha\mu$ at once from position (7, 3).

Similarly by utilizing the μ of L_{m_2} in its own row, and the λ of L_{m_1} in its own column, each non-zero asterisk term $a\mu$ situated *below* the barrier can be moved *one step diagonally upwards to the left* till it passes off the figure after reaching the first column.

Working from left to right above, and from right to left below, the barrier, we manifestly delete every asterisk.

In (ii) the original rule, involving the μ above, and the λ to the right, removes each element *one step diagonally upwards to the right* until it is deleted.

In (iii) the same rule applies, provided that each element of the first column is treated before those of the second, and each of the second before those of the third, and so on. The process differs from (ii) in that, at each remove, the lowest row receives, through the term $b_k\mu$, *an addition in the column entered by the diagonally upward moving term.* Clearance from left to right can proceed as before.

Hence any asterisk below an L_{m_k} of the matrix (28) may be removed, although the act may alter the elements of a lower asterisk, owing to column operations, but in no case can a λ term be added to an asterisk. Evidently by proceeding systematically downwards for the L columns, and similarly to the right for the L' rows, every asterisk may be removed.

Thus solely by rational equivalent operations, which can be combined with all earlier operations on Λ so as to give an equivalent matrix $P\Lambda Q$, the desired canonical form has been established. If preferred, the non-singular core S could be put into the classical form $C(\lambda, \mu)$ instead of the rational form $B(\lambda, \mu)$.

8. The Invariants of a Singular Pencil of Matrices.

Theorem II.—*The invariants of a singular pencil* Λ *under equivalent transformations are* (i) *the two sets of minimal numbers* m *and* m′ (*including of course zero values, if any,* m = 0, m′ = 0), *and* (ii) *the invariant factors of the singular core* M.

In proof of (ii), consider the co-factors in the determinant Δ of the upper right and lower left corner elements of a canonical submatrix L_m; evidently they are $\mu^m\Delta_m$ and $\lambda^m\Delta_m$, where Δ_m is a common residual minor of order $n-m-1$. The H.C.F. of these is Δ_m, so that L_m contributes nothing from itself but unity to the highest invariant factor of Λ. The same is true of the remaining L_m's and $L'_{m'}$'s, and by considering the co-factors of the upper right and lower left elements of two L_m's at once, and so on, we see that the L_m's and the $L'_{m'}$'s contribute nothing to any of the invariant factors. Thus the invariant factors reside solely in the singular core S, and if no S exists they are units alone. For example, in Ex. 2, p. 115, there is a single $m=1$, a single $m'=1$, and no invariant factor.

In regard to (i), we enunciate the following theorem:

Theorem III.—*Two pencils Λ and $\hat{\Lambda}$ are equivalent if, and only if, they possess the same minimal numbers and the same invariant factors.*

Proof.—Let m_1 be the smallest minimal number for Λ, $\hat{m}_1$ the corresponding smallest one for $\hat{\Lambda}$. If $\hat{m}_1 = m_1$, then by partial reduction to canonical form we have

$$\Lambda = H_1 \begin{bmatrix} L_{m_1} & \\ & \Lambda_1 \end{bmatrix} K_1, \quad \hat{\Lambda} = \hat{H}_1 \begin{bmatrix} L_{m_1} & \\ & \hat{\Lambda}_1 \end{bmatrix} \hat{K}_1.$$

The same argument can now be applied to the isolated matrices Λ_1 and $\hat{\Lambda}_1$, in respect of the next minimal numbers $\hat{m}_2 = m_2$, and so on, until the m's are exhausted. Further similar treatment applies to m': and the result in relation to the non-singular core is already known. Thus finally we have

$$\Lambda = HLK, \qquad\qquad \hat{\Lambda} = \hat{H}L\hat{K},$$

whence $\hat{\Lambda} = \hat{H}H^{-1}\Lambda K^{-1}\hat{K} = P\Lambda Q, \quad |P| \neq 0, \; |Q| \neq 0.$

Hence the pencils Λ and $\hat{\Lambda}$ are equivalent. The converse can be proved step by step in a similar manner.

9. Application to Singular Pencils of Bilinear Forms.

It is scarcely necessary to go into details, the matter being one of translation from one notation into another. The singular pencil of matrices $\Lambda = \lambda A_1 + \mu A_2$ being reducible to canonical form $P\Lambda Q = L$, the pencil of bilinear forms in independent variables, $y'\Lambda x$, is reducible to $\eta' L \xi$ by substitutions $y' = \eta' P$, $x = Q\xi$. The zero rows and columns in L represent redundant variables. If Λ remains singular when these are absorbed, the canonical form of the pencil consists of a sum of minimal bilinear forms such as, e.g., for $m = 2$,

$$[\eta_1, \eta_2, \eta_3] \begin{bmatrix} \lambda & \\ \mu & \lambda \\ & \mu \end{bmatrix} \begin{bmatrix} \xi_1 \\ \xi_2 \end{bmatrix} = \lambda(\eta_1\xi_1 + \eta_2\xi_2) + \mu(\eta_2\xi_1 + \eta_3\xi_2), \quad (29)$$

together with a sum of minimal forms like

$$[\eta_4, \eta_5] \begin{bmatrix} \lambda & \mu & \\ & \lambda & \mu \end{bmatrix} \begin{bmatrix} \xi_3 \\ \xi_4 \\ \xi_5 \end{bmatrix} = \lambda(\eta_4\xi_3 + \eta_5\xi_4) + \mu(\eta_4\xi_4 + \eta_5\xi_5), \quad (30)$$

(where $m' = 2$) and possibly a residual non-singular pencil in one or other of the usual canonical shapes. Evidently there are $2m + 1$ variables involved here in each such minimal form of the first kind, and $2m' + 1$ in each of the second kind. None of the variables ξ_i, η_j can be present simultaneously in two or more submatrices.

EXAMPLES

1. Singular pencils of matrices (and bilinear forms) fall into three categories, (i) those with both types m, m' of minimal number, (ii) those with m but without m', and (iii) those with m' but without m. If the pencil is generalized quadratic only case (i) can occur.

2. If $\eta' P\Lambda Q\,\xi$ is the canonical form of the pencil $y'\Lambda x$, prove that the number r of variables ξ_i is less than the number of original variables x_i by δ, the number of zeros of the minimal number m. State and prove a similar result for η, y, m'.

3. By using the canonical form, show that δ is also the number of distinct linearly independent relations, $u\Lambda = 0$, between the rows of Λ, where u is constant.

10. **Quadratic and Hermitian Pencils.**

The problem of reducing pencils of generalized quadratic forms to canonical shape requires not merely the setting up of a convenient Hermitian or symmetric matrix pencil with the same minimal numbers and invariant factors as the proposed pencil, but also the assurance that the canonical form can be obtained by a congruent or conjunctive transformation, in order that variables, which are no longer independent, may be inserted in the matrix pencils. That an *equivalent* transformation exists we know from the theory of bilinear pencils just expounded; the existence of the desired *congruent* transformation will now be deduced.

Lemma III.—*If two equivalent matrices* X *and* Y *are both respectively* *symmetric or skew symmetric, there exists a matrix* H *such that* $\mathrm{Y} = \mathrm{H'XH}$, $|\mathrm{H}| \neq 0$.

Proof.—Let $Y = P'XQ$, $|P| \neq 0$, $|Q| \neq 0$, where $X = \pm X'$, $Y = \pm Y'$.

Then $$P'XQ = Y = \pm Y' = \pm Q'X'P = Q'XP. \quad . \quad . \quad (31)$$

Hence $$(QP^{-1})'X = XQP^{-1},$$

or, let us say, $$S'X = XS, \quad S = QP^{-1}, \quad |S| \neq 0.$$

Iterating this last identity by squaring, cubing, and so on, we have

$$(S')^2X = XS^2, \ldots, (S')^kX = XS^k, \ldots, T'X = XT, \quad (32)$$

where $T = f(S)$ is a polynomial in the matrix S. If further $|T| \neq 0$, we have $X = T'XT^{-1}$ and so $Y = P'XQ = P'T'XT^{-1}Q$. This may be identified with $H'XH$, as desired, provided that $H = TP = T^{-1}Q$, or $T^2 = QP^{-1} = S$. This condition is satisfied, since S is non-singular, and T can be obtained as a polynomial in S by the interpolation formula of Chapter VI, p. 78.

Thus the Lemma is established, though in general only by resorting, at this last step, to irrationalities and the complex field. It applies to pencils in which the basic matrices are of any of the types considered, as for example the pencil in which the basis consists of a symmetric and a skew matrix. Supposing therefore that a canonical matrix pencil with the same type of basis has been constructed—it matters not how—and that it has the prescribed minimal numbers and invariant factors, then the Lemma ensures that the pencil of *forms*, as distinct from matrices, can also in all cases be reduced to canonical type.

11. Weierstrass's Canonical Pencil of Quadratic Forms.

If each isolated diagonal submatrix $C_p(\lambda, \mu)$ of the classical irrational form for a matrix pencil,

$$C_p(\lambda, \mu) = \begin{bmatrix} a\lambda + \mu & \lambda & & & \\ & a\lambda + \mu & \lambda & & \\ & & \ddots & \ddots & \\ & & & \ddots & \lambda \\ & & & & a\lambda + \mu \end{bmatrix}, \quad . \quad (33)$$

be premultiplied by the p-rowed matrix J_p of Chapter II, p. 11, we derive

$$W_p(\lambda, \mu) = \begin{bmatrix} & & & & a\lambda+\mu \\ & & & a\lambda+\mu & \lambda \\ & & \cdot^{\cdot^{\cdot}} & \cdot^{\cdot^{\cdot}} & \\ & a\lambda+\mu & \cdot^{\cdot^{\cdot}} & & \\ a\lambda+\mu & \lambda & & & \end{bmatrix}. \quad . \quad (34)$$

This clockwise rotation through a right angle can be performed for each submatrix, yielding the *symmetric* pencil

$$W(\lambda, \mu) = [W_p(\lambda, \mu)] = [J_p C_p(\lambda, \mu)] = [J_p][C_p(\lambda, \mu)],$$

which is thus a valid canonical shape for a symmetric matrix pencil with prescribed elementary divisors. Introducing variables in accordance with Lemma III, we derive the irrational canonical form for a pencil of non-singular quadratics first given by Weierstrass. It is not necessary to write the pencil of forms out in full: its nature will be inferred at once from the matrix W.

The case of Hermitian, and indeed of real quadratic, pencils requires an additional condition, called the signature test, analogous to the law of inertia of p. 89. If a and therefore the factor $a\lambda + \mu$ is complex, the invariant factor which contains it, being real (Ex. 7, p. 119), must possess $\bar{a}\lambda + \mu$ also, so that to each $W(\lambda, \mu)$ corresponds a $\overline{W}(\lambda, \mu)$, and the two can be placed astride the diagonal to form a double submatrix of the required Hermitian form,

$$\begin{bmatrix} \cdot & W(\lambda, \mu) \\ \overline{W}(\lambda, \mu) & \cdot \end{bmatrix}. \quad . \quad . \quad . \quad . \quad . \quad (35)$$

But if a is real, a submatrix $\pm W(\lambda, \mu)$ is placed upon the diagonal, where the sign must be determined by the additional test. References to proofs of this are given on p. 142.

EXAMPLES

1. Show that

$$\begin{bmatrix} \cdot & 1 & \cdot & \cdot \\ 1 & \cdot & \cdot & \cdot \\ \cdot & \cdot & \cdot & 1 \\ \cdot & \cdot & 1 & \cdot \end{bmatrix} \begin{bmatrix} a\lambda+\mu & \lambda & & \\ \cdot & a\lambda+\mu & & \\ & & b\lambda+\mu & \lambda \\ & & \cdot & b\lambda+\mu \end{bmatrix}$$

$$= \begin{bmatrix} \cdot & a\lambda+\mu & & \\ a\lambda+\mu & \lambda & & \\ & & \cdot & b\lambda+\mu \\ & & b\lambda+\mu & \lambda \end{bmatrix}.$$

The corresponding quadric pencil $\lambda S_1 = \mu S_2$ is given by

$$S_1 = 2ax_1x_2 + x_2^2 + 2bx_3x_4 + x_4^2, \quad S_2 = 2x_1x_2 + 2x_3x_4.$$

The Segre characteristic is [2, 2] if $a \neq b$, [(2, 2)] if $a = b$.

2. If in the above case $\bar{a} = b$, find the analogous premultiplier which renders this classical matrix Hermitian.

3. If two symmetric or skew symmetric matrices have the same invariant factors, the one can be transformed into the other by a unitary (or an orthogonal) transformation.

[If X and Y are the matrices, consider the characteristic pencils $\lambda X + \mu I$, $\lambda Y + \mu I$, and employ Lemma III.]

4. Show how to construct a canonical form for a pencil of skew, or symmetric, matrices with prescribed elementary divisors.

[Arrange submatrices W_p, $-W_p$ across the diagonal.]

12. Rational Canonical Form for Hermitian and Quadratic Pencils.

The Weierstrassian form just mentioned is related to the classical collineatory form C of a matrix A. We shall next show that the rational form B can be rationally transformed into symmetric shape by premultiplying by a non-singular matrix P of simple form. For example, taking an Hermitian pencil in the rational shape $B_p(\lambda, \mu)$, we have, apart from the signature test,

$$\begin{bmatrix} b_3 & b_2 & b_1 & -1 \\ b_2 & b_1 & -1 & \\ b_1 & -1 & & \\ -1 & & & \end{bmatrix} \begin{bmatrix} \lambda & \mu & & \\ & \lambda & \mu & \\ & & \lambda & \mu \\ b_4\mu & b_3\mu & b_2\mu & \lambda + b_1\mu \end{bmatrix}$$

$$= \begin{bmatrix} b_3\lambda - b_4\mu & b_2\lambda & b_1\lambda & -\lambda \\ b_2\lambda & b_1\lambda + b_2\mu & -\lambda + b_1\mu & -\mu \\ b_1\lambda & -\lambda + b_1\mu & -\mu & \\ -\lambda & -\mu & & \end{bmatrix}, \quad . \quad (36)$$

as may readily be verified in this and in the general case, with a slight variation according as p is odd or even. Since the invariant factors of an Hermitian pencil are real, the coefficients b_i are real (Ex. 7, p. 119), and so this last matrix is real symmetric and therefore Hermitian, as we desire. Simplifying it further by a succession of congruent operations, $\text{row}_{n-1} + b_1 \text{row}_n$, $\text{col}_{n-1} + b_1 \text{col}_n$, $\text{row}_{n-2} + b_2 \text{row}_n$, $\text{col}_{n-2} + b_2 \text{col}_n$, and so on; then $\text{row}_{n-2} + b_1 \text{row}_{n-1}$, $\text{col}_{n-2} + b_1 \text{col}_{n-1}$, and so on, we derive rational and real symmetrical canonical forms for Hermitian and quadratic pencils, which again differ slightly according as the order of a submatrix is odd or even, typical submatrices being, e.g.,

$$D_4(\lambda,\mu)=\begin{bmatrix} b_3\lambda-b_4\mu & & & -\lambda \\ & b_1\lambda-b_2\mu & -\lambda & -\mu \\ & -\lambda & -\mu & \\ -\lambda & -\mu & & \end{bmatrix}, \tag{37}$$

$$D_5(\lambda,\mu)=\begin{bmatrix} b_4\lambda-b_5\mu & & & & -\lambda \\ & b_2\lambda-b_3\mu & & -\lambda & -\mu \\ & & -\lambda-b_1\mu & -\mu & \\ & -\lambda & -\mu & & \\ -\lambda & -\mu & & & \end{bmatrix}. \tag{38}$$

By Lemma III, conjugate variables can be inserted, and a rational canonical shape for pencils of quadratic forms is thus established.

EXAMPLES

1. Verify that expansion of the determinant $|D_p(\lambda,\mu)|$ gives for the invariant factor the correct binary polynomial

$$B_p(\lambda,\mu)=\lambda^p+b_1\lambda^{p-1}\mu-b_2\lambda^{p-2}\mu^2+b_3\lambda^{p-3}\mu^3-\ldots+(-)^{p-1}b_p\mu^p.$$

[Expand the determinant in terms of its first row and column. This gives the recurrence relation $|B_p|=\lambda^2|B_{p-2}|-(-\mu)^{p-1}(b_{p-1}\lambda-b_p\mu)$.]

2. As before, in the corresponding canonical form for skew or skew Hermitian matrix pencils, the invariant factors occur in pairs, and submatrices D_k, $-\tilde{D}_k$ lie across the diagonal.

13. **Singular Hermitian and Quadratic Pencils.**

If the matrix pencil $\Lambda=\lambda A_1+\mu A_2=\lambda\tilde{A}_1'+\mu\tilde{A}_2'=\tilde{\Lambda}'$ is Hermitian and singular, the two sets of minimal numbers $[m]$ and $[m']$ are identical: for if $u\Lambda=0$ is a minimal relation, with minimal number m, we have by transposition the associate relation $\tilde{\Lambda}'u'=\Lambda\tilde{u}'=0$, which yields a corresponding and identical m'. Hence a canonical matrix pencil with prescribed minimal numbers can be constructed for this case, namely

$$P\Lambda Q = \begin{bmatrix} \cdot & L_{m_1} & & & & \\ L'_{m_1} & \cdot & & & & \\ & & \cdot & L_{m_2} & & \\ & & L'_{m_2} & \cdot & & \\ & & & & \ddots & \\ & & & & & M \end{bmatrix} \text{ or } \begin{bmatrix} & & & & L_{m_1} & \cdot \\ & & & L_{m_2} & & \cdot \\ & & \cdot & & & \vdots \\ & L'_{m_2} & & & & \vdots \\ L'_{m_1} & & & & & \vdots \\ \cdot & \cdots & \cdots & \cdots & \cdots & M \end{bmatrix}, \qquad (39)$$

where each L_m and L'_m is an isolated minimal submatrix, or possibly a rectangle of zeros, and M is a non-singular core, which can be cast into any desired shape. Once again inserting variables by Lemma III, we have a rational canonical shape for a singular pencil of quadratic forms. For Hermitian forms the signature test only affects M.

The analogous form for a pencil of skew or skew Hermitian matrices can also be inferred, namely

$$\begin{bmatrix} \cdot & L_{m_1} & & & & \\ -L'_{m_1} & \cdot & & & & \\ & & \cdot & L_{m_2} & & \\ & & -L'_{m_2} & \cdot & & \\ & & & & \ddots & \\ & & & & & M \end{bmatrix}. \quad . \quad . \qquad (40)$$

14. Reduction of a Pencil with a Basis of Transposed Matrices.

We consider next a pencil in which one of the basic matrices is symmetric, the other skew symmetric. Let there be a matrix pencil

$$\Lambda = \lambda A + \mu A', \quad . \quad . \quad . \quad . \quad . \quad . \quad . \qquad (41)$$

the basis thus consisting of a square matrix A and the transposed, A'. If we put $A + A' = R = R'$, and $A - A' = S = -S'$, together with

$$\lambda + \mu = 2\rho, \quad \lambda - \mu = 2\sigma, \quad . \quad . \quad . \quad . \qquad (42)$$

we have

$$\Lambda = \rho R + \sigma S, \quad . \quad . \quad . \quad . \quad . \quad . \quad . \qquad (43)$$

where R is symmetric, S skew symmetric.

Now, by transposing the determinant of the pencil, we have $\Delta = | \rho R + \sigma S | = |\rho R' + \sigma S' | = | \rho R - \sigma S |$: further, to each minor of Δ, let us say $| \rho R_k + \sigma S_k |$, there likewise corresponds a transposed minor $| \rho R_k - \sigma S_k |$ of equal value. It follows that the sequence Δ, Δ_1, Δ_2, ... consists of *even* polynomials in σ; and so the invariant factors of the pencil Λ must contain factors like $\rho + a\sigma$,

$\rho - a\sigma$, $a \neq 0$, *in pairs*. The elementary divisors of Λ must therefore be of the types ρ^p, σ^p, and $(\rho + a\sigma)^r$ paired with $(\rho - a\sigma)^r$.

By Lemma III, if suitable canonical matrices of type similar to the pencil itself can be constructed, with prescribed elementary divisors, the latter being of the nature just indicated, then a congruent transformation is available.

First, then, if the pencil Λ is singular and has a minimal number m, the minimal relation $u\Lambda = 0$ induces $\Lambda' u' = 0$, which again implies $\Lambda \hat{u}' = 0$, where $\hat{u}'$ means that $-\sigma$ has been put for σ in u'. Hence, as in the Hermitian case of the previous article, to each minimal number m there corresponds $m' = m$; so that appropriate submatrices are of the form L_m, $\hat{L}'_m$ placed symmetrically across the diagonal.

Turning next to the non-singular core, we can build up the requisite invariant factors as follows:

Case I.—Elementary divisors of type

$$(\rho + a\sigma)^r \equiv \theta^r, \quad (\rho - a\sigma)^r \equiv \hat{\theta}^r, \quad a \neq 0, \quad \theta \neq \hat{\theta}.$$

The following types of submatrix meet the requirements:

$$\begin{bmatrix} \cdot & \theta \\ \hat{\theta} & \cdot \end{bmatrix}, \quad \begin{bmatrix} & & & \theta \\ & & \theta & \hat{\theta} \\ & \hat{\theta} & & \\ \hat{\theta} & \theta & & \end{bmatrix}, \quad \begin{bmatrix} & & & & & \theta \\ & & & & \theta & \hat{\theta} \\ & & & \theta & \hat{\theta} & \\ & & \hat{\theta} & & & \\ & \hat{\theta} & \theta & & & \\ \hat{\theta} & \theta & & & & \end{bmatrix}, \text{ \&c.} \qquad (44)$$

Each of these, on inspection, has among its first minors θ^{2r-1}, $\hat{\theta}^{2r-1}$ with H.C.F. unity, so that the submatrices are elementary; they are also of the desired type $\rho R + \sigma S$.

Case II.—Elementary divisors of types ρ^p, σ^q.

Consider the specimen submatrices (where the asterisk denotes a zero):

$$[\rho], \quad \begin{bmatrix} & \rho \\ \rho & * \end{bmatrix}, \quad \begin{bmatrix} & & \rho \\ & \rho & \sigma \\ \rho & -\sigma & \end{bmatrix},$$

$$\begin{bmatrix} & & & \rho \\ & & \rho & \sigma \\ & \rho & * & \\ \rho & -\sigma & & \end{bmatrix}, \ldots, \quad \begin{bmatrix} & \sigma \\ -\sigma & \rho \end{bmatrix}, \quad \begin{bmatrix} & & & \sigma \\ & & \sigma & \rho \\ & -\sigma & \rho & \\ -\sigma & \rho & & \end{bmatrix}, \ldots . \qquad (45)$$

They are of the type $\rho R + \sigma S$, and satisfy the conditions. Those in which no asterisk appears are elementary; but those which have an asterisk, representing a zero element on the diagonal, as the skew property requires, fall into two elementary submatrices with equal elementary divisors. It will be shown that this is necessarily the case, in that *every elementary divisor of the type* $\rho^{2\nu}$ *must be accompanied by another of the same value.* This fact, and the complementary fact that *elementary divisors of the type* $\sigma^{2\nu+1}$ *must also occur in pairs,* are consequences of the following Lemma.

Lemma IV.—*A matrix with an odd number of units in the diagonal, the remaining elements being zero, cannot be equivalent to a skew matrix* $S = -S'$.

Proof.—Let M be a matrix of order n, having units for the first $2p + 1$ elements of the diagonal, the remaining elements being zero. We wish to prove that the proposed identity $PMQ = S$, $|P| \neq 0$, $|Q| \neq 0$, is impossible.

First, if M is a unit matrix I of odd order, so that $n = 2p + 1$, the Lemma is true, for the determinant on the left side of the identity would be non-zero while that of the right would be zero, since a skew symmetric determinant of odd order vanishes.

If, however, $n > 2p + 1$, consider the $(2p + 1)$th compound of the proposed identity. The compound matrix $M^{(2p+1)}$ has unity for leading element, the rest zero. Also, by Sylvester's theorem (*Invariants*, p. 87), the compounds $P^{(2p+1)}$ and $Q^{(2p+1)}$ are again non-singular, while we note that $S^{(2p+1)}$ is skew symmetric, since evidently compounds of odd class of skew symmetric matrices are skew symmetric, compounds of even class symmetric (cf. *Invariants*, p. 106). Thus the compound identity exemplifies a case, $p = 0$, of the proposed identity; and it will be sufficient to consider this case. Suppose then that

$$\begin{bmatrix} p_{11} & v \\ x & P_1 \end{bmatrix} \begin{bmatrix} 1 & \cdot \\ \cdot & \cdot \end{bmatrix} \begin{bmatrix} q_{11} & u \\ y & Q_1 \end{bmatrix} = \begin{bmatrix} \cdot & s \\ -s' & S_1 \end{bmatrix},$$

where the matrices are of order ν, say, and have been partitioned symmetrically after their leading elements, so that u, v, s are row vectors, and x, y, s' are column vectors all of order $\nu - 1$. The continued product gives

$$\begin{bmatrix} p_{11}q_{11} & p_{11}u \\ xq_{11} & xu \end{bmatrix} = \begin{bmatrix} \cdot & s \\ -s' & S_1 \end{bmatrix}.$$

Hence one or other of p_{11}, q_{11} must be zero, so that in any case both s and s' are zero and both $p_{11}u$ and xq_{11} are null vectors. Again the submatrix xu cannot be skew without being null, for we must have (cf. Chapter I, p. 4) $x_i u_i = 0$, $x_j u_j = 0$, and also $x_i u_j = -x_j u_i$, from which it follows that in any case all elements $x_i u_j$ of the submatrix xu are zero. Hence if the matrix S is skew it can only be null, and finally the null matrix cannot have odd rank and be equivalent to M. Hence the Lemma is proved.

The Lemma shows that, on putting $\rho = 0$ in any canonical form $U(\rho, \sigma)$ for the matrix pencil $\rho R + \sigma S$, it is impossible by equivalent operations to bring all the elements σ, to an *odd* number, into the diagonal. A consequence of this is that submatrices like the canonical types

$$\begin{bmatrix} \rho & \sigma & & \\ & \rho & \sigma & \\ & & \rho & \sigma \\ & & & \rho \end{bmatrix}, \quad \text{or} \quad \begin{bmatrix} \sigma & \rho & \\ & \sigma & \rho \\ & & \sigma \end{bmatrix},$$

or an odd number of such submatrices, have no equivalent of the type $\rho R + \sigma S$. The only conclusion is that pencils of this type must possess an even number, and cannot possess an odd number, of elementary divisors of the types $\rho^{2\nu}$ and $\sigma^{2\nu+1}$.

Summing up, therefore, we see that the possible types of submatrix to which such pencils $\Lambda = \lambda A + \mu A' = \rho R + \sigma S$ can be reduced by congruent transformation are the following:

$$(1)\quad \begin{bmatrix} \lambda \\ \mu \end{bmatrix}, \quad \begin{bmatrix} \lambda & \\ \mu & \lambda \\ & \mu \end{bmatrix}, \ldots, \quad [\lambda, \mu], \quad \begin{bmatrix} \lambda & \mu & \\ & \lambda & \mu \end{bmatrix}, \ldots .$$

$$(2)\quad \begin{bmatrix} & \lambda + c\mu \\ c\lambda + \mu & \end{bmatrix}, \quad \begin{bmatrix} & & & \lambda + c\mu \\ & & \lambda + c\mu & c\lambda + \mu \\ & c\lambda + \mu & & \\ c\lambda + \mu & \lambda + c\mu & & \end{bmatrix}, \ldots, \quad c \neq \pm 1.$$

$$(3)\quad [\rho], \quad \begin{bmatrix} & & \rho \\ & \rho & \sigma \\ \rho & -\sigma & \end{bmatrix}, \quad \begin{bmatrix} & & & & \rho \\ & & & \rho & \sigma \\ & & \rho & \sigma & \\ & \rho & -\sigma & & \\ \rho & -\sigma & & & \end{bmatrix}, \ldots, \quad \begin{cases} \rho = \frac{1}{2}(\lambda + \mu). \\ \sigma = \frac{1}{2}(\lambda - \mu). \end{cases}$$

$$(4)\quad \begin{bmatrix} & & & \rho \\ & & \rho & \sigma \\ & \rho & & \\ \rho & -\sigma & & \end{bmatrix}, \quad \begin{bmatrix} & & & & & \rho \\ & & & & \rho & \sigma \\ & & & \rho & \sigma & \\ & & \rho & & & \\ & \rho & -\sigma & & & \\ \rho & -\sigma & & & & \end{bmatrix}, \cdots .$$

$$(5)\quad \begin{bmatrix} & \sigma \\ -\sigma & \rho \end{bmatrix}, \quad \begin{bmatrix} & & & \sigma \\ & & \sigma & \rho \\ & -\sigma & \rho & \\ -\sigma & \rho & & \end{bmatrix}, \ldots, \text{ and}$$

$$(6)\quad \begin{bmatrix} & \sigma \\ -\sigma & \end{bmatrix}, \quad \begin{bmatrix} & & & \sigma \\ & & \sigma & \rho \\ & -\sigma & & \\ -\sigma & \rho & & \end{bmatrix}, \cdots . \qquad (46)$$

By taking $\lambda = 1$, $\mu = 0$, we derive as a special case *the canonical submatrices for a correlation,* that is, for the congruent reduction of a single matrix A, square or rectangular. The types are these:

$$(1)\quad \begin{bmatrix} 1 \\ \cdot \end{bmatrix}, \quad \begin{bmatrix} 1 & \cdot \\ \cdot & 1 \\ \cdot & \cdot \end{bmatrix}, \ldots, \quad [1, \ \cdot], \quad \begin{bmatrix} 1 & \cdot & \cdot \\ \cdot & 1 & \cdot \end{bmatrix}, \cdots ;$$

$$(2)\quad \begin{bmatrix} & 1 \\ c & \end{bmatrix}, \quad \begin{bmatrix} & & & 1 \\ & & 1 & c \\ & c & & \\ c & 1 & & \end{bmatrix}, \cdots ; \quad c \neq \pm 1.$$

$$(3)\quad [1], \quad \begin{bmatrix} & & 1 \\ & 1 & 1 \\ 1 & -1 & \end{bmatrix}, \quad \begin{bmatrix} & & & & 1 \\ & & & 1 & 1 \\ & & 1 & 1 & \\ & 1 & -1 & & \\ 1 & -1 & & & \end{bmatrix}, \cdots ;$$

$$(4)\quad \begin{bmatrix} & & & 1 \\ & & 1 & 1 \\ & 1 & & \\ 1 & -1 & & \end{bmatrix}, \cdots ; \qquad (5)\quad \begin{bmatrix} & 1 \\ -1 & 1 \end{bmatrix}, \quad \begin{bmatrix} & & & 1 \\ & & 1 & 1 \\ & -1 & 1 & \\ -1 & 1 & & \end{bmatrix}, \cdots ;$$

$$(6)\ \begin{bmatrix} & 1 \\ -1 & \end{bmatrix},\quad \begin{bmatrix} & & & & & 1 \\ & & & & 1 & 1 \\ & & & 1 & 1 & \\ & & -1 & & & \\ & -1 & 1 & & & \\ -1 & 1 & & & & \end{bmatrix}, \quad \ldots \qquad (47)$$

This completes the investigation of Chapter VII, p. 94, which went no further than obtaining a Jacobian form. The above work has been confined to the case of a symmetric-skew basis, and has not been extended to the Hermitian and skew Hermitian case, for the reason that there the argument is foiled by an essential dissimilarity of properties; for example, the determinant of a skew Hermitian matrix of odd order does not necessarily vanish identically, but is usually imaginary.

15. Rational Canonical Form of the Foregoing Pencil.

It remains to find a rational canonical form for the pencil just considered. The submatrices corresponding to the minimal numbers in the singular case are rational; and so also are those corresponding to invariant factors of the types ρ^p, σ^q, that is $(\lambda \pm \mu)^p$. There remains only the case of elementary divisors of type $(\rho \pm a\sigma)^r$, $a \neq 0$, and since these occur in pairs in any invariant factor, and since the elementary divisors $\sigma^{2\nu+1}$ also occur in pairs, we see that such an invariant factor is a homogeneous polynomial either odd or even in ρ, but certainly *even in* σ. We therefore choose two special forms of the rational canonical submatrix, $D_p(\rho, \sigma)$ or $D_p(\sigma, \rho)$, satisfying this condition, namely, e.g.,

$$D_4(\rho, \sigma) = \begin{bmatrix} b_4\rho & & & \sigma \\ & b_2\rho & \sigma & \rho \\ & \sigma & \rho & \\ \sigma & \rho & & \end{bmatrix},$$

and

$$D_5(\sigma, \rho) = \begin{bmatrix} b_4\rho & & & & -\rho \\ & b_2\rho & & -\rho & -\sigma \\ & & -\rho & -\sigma & \\ & -\rho & -\sigma & & \\ -\rho & -\sigma & & & \end{bmatrix}, \quad . \ . \ . \qquad (48)$$

and proceed to modify them rationally so that they assume the type $\rho R + \sigma S$. All that is necessary is to alter the sign of every other row, or column. Thus by rational operations we obtain the desired types symmetrical in ρ and skew in σ; for example

$$\begin{bmatrix} b_4\rho & & & \sigma \\ & -b_2\rho & -\sigma & -\rho \\ & \sigma & \rho & \\ -\sigma & -\rho & & \end{bmatrix} \text{ and } \begin{bmatrix} b_4\rho & & & & -\rho \\ & -b_2\rho & & \rho & \sigma \\ & & -\rho & -\sigma & \\ & \rho & \sigma & & \\ -\rho & -\sigma & & & \end{bmatrix}, \tag{49}$$

according as p is even or odd. By Lemma III a congruent transformation exists, not necessarily rational, which reduces the pencil $\rho R + \sigma S$ to a shape which may contain the above submatrices; and as a particular case we may supplement the list of types for the correlation by rational canonical submatrices such as

$$\begin{bmatrix} b_4 & & & 1 \\ & -b_2 & -1 & -1 \\ & 1 & 1 & \\ -1 & -1 & & \end{bmatrix} \text{ and } \begin{bmatrix} b_4 & & & & -1 \\ & -b_2 & & 1 & 1 \\ & & -1 & -1 & \\ & 1 & 1 & & \\ -1 & -1 & & & \end{bmatrix}. \tag{50}$$

EXAMPLE

Examine whether a rational canonical submatrix can be constructed for a pencil based on an Hermitian and a skew Hermitian matrix.

[In the invariant factors an elementary divisor of type $(\rho + \alpha\sigma)^r$ must necessarily be accompanied, if $\alpha \neq \bar{\alpha}$, by another of type $(\rho - \bar{\alpha}\sigma)^r$. This will cause the coefficients b_i to be alternately real and imaginary.]

16. **Historical Note.**—The irrational canonical form for a pencil of quadratics was given by Weierstrass in a classical memoir, *Berlin Monatsb.* (1868), 310, or *Werke*, II, 19. The existence of singular pencils was mentioned by him but excluded from discussion. The discussion of the singular case is due to Kronecker, *Berlin Sitzungsb.* (1890), 1375, and (1891), 9, 33. A treatment by rational methods has recently been given by Dickson, *Trans. Am. Math. Soc.*, **29** (1927), 239–53. The Hermitian analogue of the quadratic pencil of Weierstrass has been given by M. I. Logsdon, *Am. Jour. Math.*, **44** (1922), 254, but the signature test is not investigated.

A rational canonical pencil for Hermitian and quadratic forms, somewhat different from that of the present chapter, appears in Dickson's *Modern Algebraic Theories*, Chicago, 1926, 127–8.

The theorem that in a pencil based on a symmetric and a skew symmetric matrix elementary divisors of the types $\rho^{2\nu}$, $\sigma^{2\nu+1}$ must occur in pairs was given by Kronecker, *Berlin Monatsb.* (1874), 397, or *Werke*, I, 423. Later proofs are due to Stickelberger, *J. für Math.*, **86** (1879), 42–3, and Frobenius, *Encyc. des Sc., Math.*, I, **2**, 463–9. They are connected with generating series for the reciprocal matrix pencil and incidentally with the methods employed by Darboux for the reduction of a pencil of quadratic forms (to which we shall briefly allude in the next chapter).

The actual reduction of a singular pencil, for the general cases of a collineation and of a correlation, is still outstanding.

Concerning the description and enumeration of types of pencils of conics and quadrics we may refer to Bromwich's Tract (Cambridge, 1906), 46–7, 60; or Hilton, *Linear Substitutions* (Oxford, 1914), 104, 105. The simple enumeration of types depends on the Segre characteristic and therefore on the theory of partitions in combinatory analysis. Generating functions can be constructed for this enumeration; see Bromwich, p. 60. See also Bromwich, p. 69, for theorems concerning the *signature* of real quadratic pencils and of Hermitian pencils—in particular for a proof of a theorem on the signature, due to Klein. Cf. also Bromwich, *Proc. London Math. Soc.* (1), **32** (1900), 349. The signature test is established by P. Muth, *J. für Math.*, **128** (1905), 302–21, for real equivalence of real non-singular quadratic pencils, and by H. W. Turnbull, *Proc. London Math. Soc.* (2), **39** (1935), 232–48, for Hermitian and singular pencils; also G. R. Trott, *Am. J. Math.*, **56** (1934), 359–71. Other treatments have been given by M. H. Ingraham and K. W. Wegner, *Trans. Amer. Math. Soc.*, **38** (1935), 145–62, J. Williamson, *Am. J. Math.*, **57** (1935), 475–90. Further writings on singular pencils are by A. C. Aitken, *Q. J. Math.* (Oxford), **4** (1933), 241–5, H. W. Turnbull, *Proc. Edin. Math. Soc.*, Ser. II, **4** (1934), 67–76, W. Ledermann, *Proc. Edin. Math. Soc.*, Ser. II, **4** (1934), 92–105.

A paper treating of combinants based on three matrices, by O. E. Brown, *Bull. Amer. Math. Soc.*, **37** (1931), 424–6, cites earlier work by S. Kantor, *Sitzgb. Akad. Münch.*, **27** (1897), and *Monatsh. f. Math. u. Phys.*, **11** (1900); also G. E. F. Sherwood, *Thesis* (Chicago, 1922).

CHAPTER X

Applications of Canonical Forms to Solution of Linear Matrix Equations. Commutants and Invariants

1. The Auxiliary Unit Matrices.

In non-commutative algebra the equation $AX = XA$ is no triviality, but possesses an interest which has lately been reinforced by an application to quantum algebra. To effect a general solution—that is, *to find the most general matrix X which commutes with a given matrix A*—we introduce the reader to a few further properties of the auxiliary unit matrix U (p. 62).

The immediate action of U and U' upon any matrix $X = [x_{ij}]$ with which they conform is that of *translating the rows or columns, one place up or down, left or right.* We have in fact,

$$\left.\begin{aligned} UX &= \begin{bmatrix} x_{i+1,j} \\ \cdot \end{bmatrix}, \qquad U'X = \begin{bmatrix} \cdot \\ x_{i-1,j} \end{bmatrix} \\ XU' &= [x_{i,j+1}, \quad \cdot], \quad XU = [\cdot, \quad x_{i,j-1}] \end{aligned}\right\} \quad . \quad . \quad . \quad (1)$$

The notation means, for example, that the ijth element of UX is $x_{i+1,j}$ where the original first row has disappeared and the lowest row of the product matrix is zero. Similarly for the others. These four operations are so useful in practice that it is convenient to signalize their effects as follows:

$$\left.\begin{matrix} UX & \text{up,} \\ U'X & \text{down,} \\ XU' & \text{left,} \\ XU & \text{right.} \end{matrix}\right\} \quad . \quad . \quad . \quad . \quad . \quad . \quad . \quad . \quad (2)$$

The matrix X may, of course, be rectangular.

Lemma I.—*The general solution of* $UX = XU$ *is*

$$X = x_0 I + x_1 U + x_2 U^2 + \ldots + x_{n-1} U^{n-1} \equiv f(U), \qquad (3)$$

namely, a polynomial involving n *arbitrary parameters* x_i, *where* U *is the auxiliary unit matrix of order* n.

Proof.—If $X = [x_{ij}]$ we have

$$UX = \begin{bmatrix} x_{21} & x_{22} & \dots \\ x_{31} & x_{32} & \dots \\ \dots & \dots & \dots \\ x_{n1} & x_{n2} & \dots \\ . & . & \dots \end{bmatrix} = \begin{bmatrix} . & x_{11} & x_{12} & \dots \\ . & x_{21} & x_{22} & \dots \\ \dots & \dots & \dots & \dots \\ . & x_{n1} & x_{n2} & \dots \end{bmatrix} = XU.$$

Equating elements standing at the ijth position of these matrices

$$x_{i+1,j} = x_{i,j-1}, \quad 0 < i < n, \quad 1 < j < n+1,$$

and $x_{21} = x_{31} = \dots = x_{n1} = x_{n2} = \dots = x_{n,n-1} = 0$. On writing x_k for $x_{1,k+1}$ we conclude that

$$X = \begin{bmatrix} x_0 & x_1 & x_2 & \dots & x_{n-1} \\ . & x_0 & x_1 & \dots & x_{n-2} \\ . & . & x_0 & \dots & x_{n-3} \\ \dots & \dots & \dots & \dots & \dots \\ . & . & . & \dots & x_0 \end{bmatrix}. \quad . \quad . \quad . \quad . \quad . \quad (4)$$

The equation $UX = XU$ expresses the fact that consecutive, and therefore all, elements of the principal diagonal, or of a parallel diagonal, are equal. On expanding X in the usual way (p. 62) we have the above polynomial $f(U)$; and the Lemma is proved.

Conversely, it is evident that $f(U)$ commutes with U, though it is not at once apparent that this gives the general solution. Again, the series $\sum_{r=1}^{m} x_r U^r$, $\sum_{r=1}^{\infty} x_r U^r$ are evidently solutions: but they are no more general than the series already given, since $U^r = 0$ when $r \geqslant n$. The parameters x_r $(r \geqslant n)$ are in fact illusory.

Corollary.—*The general solution of* $U'X = XU'$ *is* $X = f(U')$.

Lemma II.—*If* U *and* V *are auxiliary unit matrices of orders* r *and* s *respectively, the general solution of* $UX = XV$ *is the rectangular* $r \times s$ *matrix*

$$X = f(U)I_{rs} = I_{rs}f(V), \quad . \quad . \quad . \quad . \quad . \quad (5)$$

where I_{rs} *is the unit matrix of* r *rows and* s *columns.*

Proof.—By I_{rs} we mean

$$I_{rs} = [\cdot,\ I_r] \text{ if } r < s, \quad I_{rs} = \begin{bmatrix} I_s \\ \cdot \end{bmatrix} \text{ if } r > s. \quad . \quad (6)$$

The proof is similar to that of Lemma I. For X to conform with U and V it must have r rows and s columns. The number of arbitrary parameters in $f(U)$ and in $f(V)$ is equal to the smaller of r and s: all others are illusory.

EXAMPLE

If $r = 4$, $s = 3$, the general solution of the equation

$$\begin{bmatrix} \cdot & 1 & \cdot & \cdot \\ \cdot & \cdot & 1 & \cdot \\ \cdot & \cdot & \cdot & 1 \\ \cdot & \cdot & \cdot & \cdot \end{bmatrix} X = X \begin{bmatrix} \cdot & 1 & \cdot \\ \cdot & \cdot & 1 \\ \cdot & \cdot & \cdot \end{bmatrix} \text{ is } X = \begin{bmatrix} x_0 & x_1 & x_2 \\ \cdot & x_0 & x_1 \\ \cdot & \cdot & x_0 \\ \cdot & \cdot & \cdot \end{bmatrix}.$$

This can be written alternatively, to illustrate (5), as

$$X = \begin{bmatrix} 1 & \cdot & \cdot \\ \cdot & 1 & \cdot \\ \cdot & \cdot & 1 \\ \cdot & \cdot & \cdot \end{bmatrix} \begin{bmatrix} x_0 & x_1 & x_2 \\ \cdot & x_0 & x_1 \\ \cdot & \cdot & x_0 \end{bmatrix} = \begin{bmatrix} x_0 & x_1 & x_2 & x_3 \\ \cdot & x_0 & x_1 & x_2 \\ \cdot & \cdot & x_0 & x_1 \\ \cdot & \cdot & \cdot & x_0 \end{bmatrix} \begin{bmatrix} 1 & \cdot & \cdot \\ \cdot & 1 & \cdot \\ \cdot & \cdot & 1 \\ \cdot & \cdot & \cdot \end{bmatrix}.$$

Here $f(U) = x_0 I + x_1 U + x_2 U^2 + x_3 U^3$, so that $f(U)$ is this four-rowed square matrix, whereas $f(V)$ is the three-rowed square matrix.

If, however, $r = 3$, $s = 4$, then

$$X = \begin{bmatrix} \cdot & x_0 & x_1 & x_2 \\ \cdot & \cdot & x_0 & x_1 \\ \cdot & \cdot & \cdot & x_0 \end{bmatrix} = \begin{bmatrix} \cdot & 1 & \cdot & \cdot \\ \cdot & \cdot & 1 & \cdot \\ \cdot & \cdot & \cdot & 1 \end{bmatrix} \begin{bmatrix} x_0 & x_1 & x_2 & x_3 \\ \cdot & x_0 & x_1 & x_2 \\ \cdot & \cdot & x_0 & x_1 \\ \cdot & \cdot & \cdot & x_0 \end{bmatrix}.$$

In these two examples the number of arbitrary parameters is three. In fact x_0, x_1, x_2 are the necessary parameters, while x_3 is illusory.

Lemma III.—*If* $\alpha \neq 0$ *and* α, β, γ *are scalar, the equation* $\alpha X + \beta UX + \gamma XV = 0$ *has only a null solution* $X = 0$.

Proof.—The matrix equation gives

$$\alpha x_{ij} + \beta x_{i+1,j} + \gamma x_{i,j-1} = 0.$$

If $\alpha \neq 0$, the elements x_{ij}, read from left to right along the bottom row, then on the next higher row, and so on, are all found to vanish: and hence $X = 0$.

Problem I.—*To solve* AX = XA.

Solution.—Reduce A, which must be square, to classical form

$$HAH^{-1} = [aI_p + U] = C, \quad . \quad . \quad . \quad . \quad . \quad (7)$$

so that the given equation becomes $H^{-1}CHX = XH^{-1}CH$: or $CY = YC$, where $Y = HXH^{-1}$. By finding a general solution Y for the latter equation, we can at once determine a general solution $X = H^{-1}YH$ of the former.

Now C is symmetrically partitioned (p. 6) into submatrices $C_p(a)$, $C_q(\beta)$, . . . : let Y be partitioned into $[Y_{ij}]$ in exactly the same way, so that Y_{11} is a square submatrix of order p, and Y_{12} is rectangular and of orders $p \times q$, and so on. For example, if there are two-by-two partitions, the equation $CY = YC$ becomes

$$\begin{bmatrix} aI_p+U, & \\ & \beta I_q+V \end{bmatrix} \begin{bmatrix} Y_{11} & Y_{12} \\ Y_{21} & Y_{22} \end{bmatrix} = \begin{bmatrix} Y_{11} & Y_{12} \\ Y_{21} & Y_{22} \end{bmatrix} \begin{bmatrix} aI_p+U, & \\ & \beta I_q+V \end{bmatrix}.$$

Hence, equating the appropriate submatrices, we have

$$(aI_p + U)Y_{11} = Y_{11}(aI_p + U), \quad (aI_p + U)Y_{12} = Y_{12}(\beta I_q + V), \quad (8)$$

and so on. These two equations are typical of the case in general, when there are $\nu \times \nu$ partitions. On cancelling out identically the first term on each side of the first equation, we have

$$UY_{11} = Y_{11}U,$$

and the values of Y_{ii} are given at once by Lemma I.

Similarly, from the second,

$$(a - \beta)Y_{12} + UY_{12} - Y_{12}V = 0,$$

where, if $a = \beta$, Lemma II applies; and if $a \neq \beta$, then $Y_{12} = 0$ by Lemma III. Similarly for each partition.

A submatrix Y_{ij} which is aligned with two distinct latent roots of A, upon the diagonal, is accordingly zero: aligned with two equal latent roots it is triangular and of the type already under view. We have therefore obtained a general solution Y which can be epitomized by writing

$$Y = [I_{r_i s_j} f_{ij}(U_j)] = [f_{ij}(U_i) I_{r_i s_j}], \quad . \quad . \quad . \quad (9)$$

where $f_{ij}(\lambda)$ is an arbitrary polynomial, unless $i \neq j$, $a \neq \beta$ when it is zero.

EXAMPLES

1. If $AX = XA$, and $HAH^{-1} = C = \operatorname{diag}(C_3(\alpha), C_2(\alpha), C_2(\beta), C_1(\beta))$, $\alpha \neq \beta$, then the Segre characteristic is $\begin{bmatrix} 3 & 2 \\ 2 & 1 \end{bmatrix}$, and we shall have

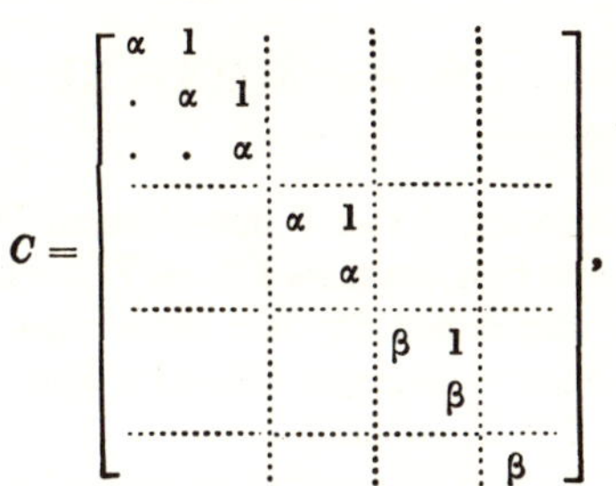

$$H^{-1}YH = X = H^{-1}\left[\begin{array}{ccc:cc:cc:c} x_0 & x_1 & x_2 & y_0 & y_1 & & & \\ & x_0 & x_1 & & y_0 & & & \\ & & x_0 & & & & & \\ \hdashline & z_0 & z_1 & t_0 & t_1 & & & \\ & & z_0 & & t_0 & & & \\ \hdashline & & & & & \xi_0 & \xi_1 & \eta_0 \\ & & & & & & \xi_0 & \\ \hdashline & & & & & & \zeta_1 & \tau_0 \end{array}\right]H,$$

where $x_0, x_1, \ldots, \tau_0$ are 14 arbitrary parameters.

2. If the Segre characteristic of A is $[p_{ij}]$, the number of arbitrary parameters in Y is

$$p_{11} + 3p_{21} + 5p_{31} + \ldots + p_{12} + 3p_{22} + 5p_{32} + \ldots + \ldots,$$

where $p_{11} \geqslant p_{21} \geqslant \ldots \geqslant p_{\nu 1}$; $p_{12} \geqslant p_{22} \geqslant \ldots$; ....

[Count the number in each submatrix of Y.]

3. Write down the form of Y when $\alpha = \beta$ above; and also when the Segre characteristic is $\begin{bmatrix} 3 & 2 & 1 \\ 2 & & \end{bmatrix}$.

3. Verify the working rule: To form Y, fill each non-zero submatrix by diagonals, starting from each right-hand top corner, and continuing until a row or a column is completely filled.

2. Commutants.

When $AX = XB$, X is called a commutant of A and B; and it is sometimes written $X = (A, B)$. Evidently we have already found the commutant (A, A): and the more general case (A, B) will follow by the same methods. It will appear that, unless A and B have a common latent root, X can only be zero. If A and B are rectangular, we shall suppose that just enough zero rows or columns are adjoined to either, to make them square; but they need not be of the same order.

Problem II.—*To solve the equation* AX = XB, *when* A *and* B *are given square matrices.*

Solution.—On throwing A and B into classical forms $A = H^{-1}CH$, $B = K^{-1}DK$, our equation becomes $CY = YD$, where $Y = HXK^{-1}$. Let $Y = [Y_{ij}]$ be partitioned by horizontal lines to conform with C and by vertical lines to conform with D. The preceding methods will then apply: and if α_i is the latent root of C associated with its ith submatrix, and if β_j is the corresponding latent root of D, then $Y_{ij} \neq 0$ only if $\alpha_i = \beta_j$.

EXAMPLES

1. If $C = \text{diag}(C_3(\alpha),\ C_2(\alpha),\ C_1(\beta))$, $D = \text{diag}(C_2(\alpha),\ C_2(\beta),\ C_2(\beta),\ C_1(\gamma))$, where α, β, γ all differ, then

$$Y = \left[\begin{array}{cc|cc|cc|c} x_0 & x_1 & & & & & \\ \cdot & x_0 & & & & & \\ \cdot & \cdot & & & & & \\ \hline y_0 & y_1 & & & & & \\ \cdot & y_0 & & & & & \\ \hline \cdot & \cdot & \cdot & z_0 & \cdot & t_0 & \cdot \end{array}\right].$$

The non-zero submatrices involving x and y correspond to the *two* α-submatrices of C and the *one* α-submatrix of D. Of the three partitions by rows, the upper pair refer to α and the lowest to β: of the four partitions by columns, the first refers to α, the next pair to β, and the last to γ.

2. Prove that a commutant (A, B) is also a commutant $(f(A), f(B))$, where $f(A)$ is a scalar function of A.

3. Prove that the mth compound of a commutant (A, B), of two square matrices of equal orders, is equal to a commutant of their mth compounds.

4. If, and only if, A and B are equivalent in the collineatory group, they have a non-singular commutant.

The theory of commutants is virtually a theory of *partial collineatory equivalence*. Matrices which have no commutant have no common latent root, and may be called totally unequivalent: those which have commutants must have some common latent root, or roots; and such may be called partially equivalent. Those which have identical latent roots and identical Segre characteristics are equivalent in our original sense.

It is clear that the solutions to Problems I and II are not quite final, in that the problems must, by their nature, possess *rational* solutions, whereas the classical forms which have been utilized are irrational. We leave to the reader the consideration of the rational solution.

3. Scalar Function of a Matrix.

Problem III.—*To find the general matrix* X *which commutes with every* P *which in turn commutes with a given* A.

Solution.—We require a matrix X such that $PX = XP$ whenever $PA = AP$. Reduce A to classical form $H^{-1}CH$, taking $Y = HXH^{-1}$, $Q = HPH^{-1}$. Then we require a Y such that $QY = YQ$ whenever $QC = CQ$.

If, as before, $C = [C_{p_i}(a_i)]$, $(i = 1, 2, \ldots, \nu)$, then the matrix $[C_{p_i}(\beta_i)]$ commutes with C for all values of the β_i, and is therefore a particular Q. Taking ν distinct values β_i, we infer, by Problem I, that the general Y which commutes with this Q is

$$Y = \operatorname{diag}(f_{p_1}(U), f_{p_2}(V), \ldots) \quad . \quad . \quad . \quad . \quad (10)$$

containing again ν diagonal submatrices conforming with those of C. If all ν latent roots a_i are distinct, this is the solution. But if two or more are equal, let us say $a_1 = a_2$, then a possible Q which commutes with C is

$$Q = \begin{bmatrix} C_{p_1}(a) & I_{p_1p_2} & \cdot & \\ \cdot & C_{p_2}(a) & \cdot & \\ \cdot & \cdot & C_{p_3}(a_3) & \\ & & & \ddots \end{bmatrix}, \quad a = a_1 = a_2.$$

But it follows at once from the condition $QY = YQ$ that

$$f_{p_1}(U) I_{p_1p_2} = I_{p_1p_2} f_{p_2}(V),$$

or simply $f_{p_1} = f_{p_2}$. Hence the arbitrary parameters of f_{p_i} can only be distinct from those of f_{p_j} if $a_i \neq a_j$; so that the number of parameters included in Y is at most p, the sum of the orders p_{ij} of the submatrices of highest order for each distinct latent root. In other words p is the order of the R.C.F. of A (or C).

But this Y will evidently commute with every commutant (C, C), and hence affords a general solution of the problem.

EXAMPLES

1. If $C = \text{diag}(C_3(\alpha), C_2(\alpha), C_2(\beta))$, $\alpha \neq \beta$, then

$$Y = \text{diag}\left(\begin{bmatrix} x_0 & x_1 & x_2 \\ & x_0 & x_1 \\ & & x_0 \end{bmatrix}, \begin{bmatrix} x_0 & x_1 \\ & x_0 \end{bmatrix}, \begin{bmatrix} y_0 & y_1 \\ & y_0 \end{bmatrix}\right)$$

containing five arbitrary parameters.

2. Find the value of Y if $\alpha = \beta$, showing that it has three parameters.

4. Connexion between Matrix Functions and Quantum Algebra.

If by a regular scalar function of A we mean a function $f(A)$ (p. 73) capable of polynomial form $\sum_{i=0}^{p-1} a_i A^i$, then we may prove the following theorem.

Theorem I.—*Every matrix* X, *which commutes with every* P *which in turn commutes with* A, *is a regular scalar function of* A. *Conversely if* $X = f(A)$, *then* $PX = XP$ *whenever* $PA = AP$.

Proof.—The converse is verifiable at once. The direct theorem follows by proving its truth for Y, when A is thrown into classical form C. But by taking the above form Y, with its p parameters, and assuming that Y may be expressed as a polynomial $\sum_{i=0}^{p-1} a_i C^i$, it follows, on comparing typical elements, that there are exactly p linear equations for the a_i in terms of the parameters and the latent roots of C: and that these equations are compatible and uniquely soluble, since their determinant is the non-zero confluent alternant belonging to the chief invariant factor of C.

For instance, in Ex. 1 above, the alternant is $\Delta(\alpha\alpha\alpha\beta\beta)$. This proves the theorem.

Thus we have two distinct, but entirely equivalent, definitions of a scalar function of a matrix—the original, and now this new definition, which incidentally has the advantage of immediate extension to infinite matrices and to other non-commutative entities. It is, in fact, taken by Dirac as the starting-point of a function theory in quantum algebra. We have space here only to hint at the interesting investigation which can be carried out merely by invoking this catalytic agent $\tilde{P}$ in order to establish a functional relation between A and X, and without recourse to the latent roots of A.

EXAMPLES

1. If X is a function of Y, then Y is a function of X.

2. If X is a function of Y, and Y is a function of Z, then X is a function of Z.

5. Scalar Functions of Two Matrix Variables.

Problem IV.—*To express a scalar function of any number of square matrices of order* n, *as a function of two matrices.*

Solution.—Consider the identity

$$A = [a_{ij}] = \sum_{i,j=1}^{n} a_{ij} U^{n-i} U'^{n-1} U^{j-1}, \quad . \quad . \quad . \quad (11)$$

where U and U' are the auxiliary unit matrices of § 1, the identity following immediately from the properties there cited. Here we have expressed the matrix A as a sum of n^2 matrices, one for each element a_{ij}: and each term of the sum is a matrix containing the single element a_{ij} in its proper position in an array otherwise null.

By treating each argument A which enters into a scalar function $\phi(A, B, C, \ldots)$ of several matrices in this way, we evidently reduce it to a scalar function $\psi(U, U')$ of two matrices only.

It will be noticed that U' appears to the same power $(n - 1)$ in each term—a useful but not an imperative restriction: for there are manifestly many alternative ways of expressing A as a series in U and U'. Besides, there are other ways of choosing a pair of matrices in terms of which all square matrices of the same order can be expressed. For instance, if $J = [\delta_{i,\, n+1-j}]$ is the secondary unit matrix of p. 11 (expressed in the Kronecker delta notation), then in general it is easy to verify that

$$JU = U'J, \quad U' = JUJ, \quad U = JU'J. \quad . \quad . \quad (12)$$

Hence any function $\psi(U, U')$ can at once be expressed as a function either of U and J, or of U' and J, where no longer both, but one only of the matrix arguments is singular.

EXAMPLES

1. Show that $U^r U'^r + U'^{n-r-1} U^{n-r-1} = I$.

2. Show that $f(U')J = Jf(U)$, $f(U)J = Jf(U')$.

3. The operation of J can be described as follows: JA reflects *rows* of A about a horizontal bisector: AJ reflects *columns* of A about a vertical bisector.

4. Solve the equations (i) $UX = \pm X'U$.

(ii) $UX = \pm X'V$.

(iii) $AX = X'B$.

5. Solve the equation $XAX = B$.

[Find, by the Sylvester formula, p. 76 or otherwise, the square root of BA. Then, $X = A^{-1}\sqrt{AB} = \sqrt{BA}\,.\,A^{-1}$. What restrictions govern A and B?]

6. Symmetric Matrices and Resolution into Factors.

Problem V.—*To resolve a given square matrix* A *into two symmetrical factors.*

Solution.—It is only necessary to resolve the canonical form $HAH^{-1} = B$ into two symmetric factors C, D; for we may then take

$$A = PQ, \quad \text{where } P = H^{-1}CH'^{-1},\ Q = H'DH, \quad . \quad (13)$$

giving the symmetric factors P and Q, consistent with the equation $B = CD$.

This resolution of B in turn is effected by a similar resolution of each submatrix $C_k(a)$ of B, an operation which was in fact performed on p. 132, when the Weierstrassian canonical form of a pencil was considered. This at once gives a solution

$$B = [J_k].[J_k]\,[C_k(a)] = C.D, \quad C = [J_k] = C', \quad D = D', \quad (14)$$

where $C_k(a)$ is a typical submatrix of the classical type, J_k being the corresponding secondary unit matrix, so that $[J_k]^2 = [I_k] = I$. The symmetry of $D = D'$ follows from Ex. 2 above, in that $C_k(a)$ is a simple linear instance of the polynomial $f(U)$.

Manifestly there are a great many alternative ways of performing this resolution of B.

EXAMPLES

1. $$\begin{bmatrix} \alpha & 1 & & & \\ & \alpha & 1 & & \\ & & \alpha & & \\ & & & \beta & 1 \\ & & & & \beta \end{bmatrix} = \begin{bmatrix} \cdot & \cdot & 1 & & \\ \cdot & 1 & \cdot & & \\ 1 & \cdot & \cdot & & \\ & & & \cdot & 1 \\ & & & 1 & \cdot \end{bmatrix} \cdot \begin{bmatrix} \cdot & \cdot & \alpha & & \\ \cdot & \alpha & 1 & & \\ \alpha & 1 & \cdot & & \\ & & & \cdot & \beta \\ & & & \beta & 1 \end{bmatrix} = \begin{bmatrix} J_3 & \\ & J_2 \end{bmatrix} . D.$$

2. If $f(U)$ is an arbitrary non-singular polynomial in U, find a unique polynomial $g(U)$ such that $f(U)g(U) = \alpha I + U$.

3. In the symmetric factors C, D either the first or the second submatrix whose product is $C_k(\alpha)$ may be derived from an arbitrary non-singular polynomial $f_k(U)$.

[Either take $C = [f_k(U)][J_k]$, $D = [J_k][g_k(U)]$, or the same with f and g interchanged. Then $C = C'$, $D = D'$, $|C|$ or $|D| \neq 0$.]

4. *A collineation may always be resolved into successive polar reciprocations with regard to two quadrics* S *and* Σ', *at least one of which is non-singular.*

[The polar of a point x in a quadric S has a pole y with regard to a second quadric Σ'. The two quadrics, in this order, set up a collineation A. Conversely, by Problem V, any collineation A has two symmetric factors P, Q, which determine two such quadrics.

If $x'Qx = 0$ and $uPu' = 0$ are the point and tangential equations of these quadrics, x and u being contragredient vectors—point and prime vectors respectively—then the *polar* of the point x in Q is the prime-vector $x'Q$, while the *pole* of the prime u in P is the point-vector Pu'.

Successive reciprocation gives $u = x'Q$, $y = Pu'$, so that $y = PQx$, if we remember that $Q = Q'$. Hence $y = Ax$.]

5. What geometrical connexion is there between the latent points of the collineation and these two quadrics?

[In general, the n vertices of the common self-conjugate simplex of the quadrics are evidently latent for the double reciprocation, and therefore coincide with the n latent points. This gives an *a priori* reason in favour of this algebraic theorem.]

6. If $|P| \neq 0$, show that the pencil $\lambda P^{-1} + \mu Q$ is equivalent to $\lambda I + \mu A$: and that the collineatory reduction of A to classical form induces a congruent reduction of this pencil.

7. Investigate the possibility of resolving a given A into two factors P, Q which are symmetric, Hermitian, or skew.

[Restrictions such as those holding for the pencil $\lambda A + \mu A'$ on p. 135 must be imposed. P and Q can both be Hermitian, if, and only if, to each $C_k(\alpha)$ there corresponds a $C_k(-\alpha)$ when α is complex.

If P and Q are respectively symmetric and skew symmetric, then there must be submatrices $C_k(\alpha)$, $C_k(-\alpha)$ occurring in pairs, for each case $\alpha \neq 0$.

If both P and Q are skew symmetric, pairs $C_k(\alpha)$, $C_k(\alpha)$ must occur, for each α.]

7. Invariants or Latent Forms of a Matrix.

We shall now give a brief account of latent forms, enough to show the scope of the theory, and how it illustrates the solution of matrix equations involving the auxiliaries U and U'. Just as an orthogonal transformation $\xi = Ax$, $(A'A = I)$, leaves a certain quadric $x'Ix = \xi'I\xi$ unchanged, so a more general matrix is capable of possessing similar properties.

Suppose that $f(x)$ is a polynomial in n variables x, then $f(x)$ is said to be an *absolute invariant* of the transformation $\xi = Ax$, if, and only if, the following identity holds for all values of x:

$$f(\xi) \equiv f(Ax) = f(x) \neq 0, \ \{x\}. \quad . \quad . \quad . \quad (15)$$

If $f(Ax) = \phi \, . f(x) \neq 0, \{x\}$, where ϕ is a scalar expression independent of x, then f is said to be a *relative invariant*. In either case $f(x)$ is said to be *latent* in the transformation.

The several non-homogeneous parts of a polynomial invariant are themselves invariants, as is seen by changing x to ρx in (15), where ρ is a non-zero scalar. Hence we may confine our attention to homogeneous forms. Again, if $f(x)$ is an invariant of A then $f(Px)$ is an invariant of $B = P^{-1}AP$, $(|P| \neq 0)$. For we may write Px for x in the identity and obtain $f(APx) = \phi f(Px)$. But $AP = PB$: so that $f(PBx) = \phi f(Px)$ for all x. Hence $f(Px)$ is an invariant of B.

It must be noticed that $f(Ax)$ is not a scalar function of a matrix, but is actually a scalar function $f(\xi_1, \ldots, \xi_n)$ of n arguments, written in the contracted notation.

EXAMPLES

1. Show that the linear invariants of A are given by the latent primes of the collineation A.

[If $\Sigma v_i x_i$ is the linear invariant, where v denotes the row-vector of its n coefficients, then by definition $vAx = \varphi vx$. Hence, if $v \neq 0$, φ is equal to a latent root α.

The general values of v are given by $u\psi(A)/(A - \alpha I)$, where u is a vector of maximum grade p, $\psi(A)$ is the R.C.F. of A, and where α is any one of the ν distinct latent roots (p. 71).]

2. Absolute linear invariants exist if, and only if, there is a latent root equal to unity.

3. If $x'Qx$ is a latent quadratic form for A, then $A'QA = \phi Q$.

[This follows directly from the definition.]

4. Adapt the concept of invariance to the case of several sets of variables: more especially to contragredient sets u, x.

$$[f(\theta, \xi) \equiv f(uA^{-1}, Ax) = \varphi f(u, x) \not\equiv 0, \quad \{u, x\}.]$$

5. If uBx is a latent bilinear form, for A, when $\xi = Ax$, $\theta = uA^{-1}$ are contragredient transformations, then $A^{-1}BA = \varphi B$. Hence the problem of discovering latent bilinear forms for a collineation A is essentially a problem of commutants.

6. The general solution of $UX = \alpha XU$ is

$$X = H_\alpha f(U),$$

where $$H_\alpha = \text{diag}\,(1, \alpha, \alpha^2, \ldots, \alpha^{n-1}).$$

7. $$\begin{bmatrix} 1 & & \\ & \alpha^{-1} & \\ & & \alpha^{-2} \end{bmatrix} \begin{bmatrix} \alpha & 1 & \\ & \alpha & 1 \\ & & \alpha \end{bmatrix} \begin{bmatrix} 1 & & \\ & \alpha & \\ & & \alpha^2 \end{bmatrix} = \begin{bmatrix} \alpha & \alpha & \\ & \alpha & \alpha \\ & & \alpha \end{bmatrix} = \alpha \begin{bmatrix} 1 & 1 & \\ & 1 & 1 \\ & & 1 \end{bmatrix}; \; \alpha \neq 0.$$

8. **Latent Quadratic Forms.**

By Ex. 3 above, if $x'Qx$ is a quadratic form latent in a transformation A, then

$$A'QA = \phi Q, \quad Q = Q'. \quad . \quad . \quad . \quad . \quad . \quad (16)$$

It is possible to discover a general solution Q by throwing A into classical form, just as in Problem I, p. 146. In fact if $C = HAH^{-1}$ and $Q = H'YH$, then

$$C'YC = \phi Y, \quad Y = Y': \quad . \quad . \quad . \quad . \quad . \quad (17)$$

and the problem is reduced to the canonical case. To solve this we shall first consider the following equation

$$aX + U'X + XV + U'XV = 0, \quad . \quad . \quad . \quad (18)$$

where a is scalar. As in Lemma III, p. 145, on considering elements x_{ij} of X in the order $\text{row}_1 \rightarrow$, $\text{row}_2 \rightarrow, \ldots,$ it follows that

$$\text{if } a \neq 0, \text{ then } X = 0. \quad . \quad . \quad . \quad . \quad . \quad (19)$$

If $a = 0$, the equation can be written

$$U'X + XV + U'XV = 0, \text{ or } U'X = X\Theta, \quad \Theta = -V(I + V)^{-1}. \quad (20)$$

The most convenient way to solve this is as follows:

Let three adjacent elements x_{ij}, $x_{i,\,j+1}$, $x_{i+1,\,j}$ be enclosed in a gnomon: then the sum of these three elements must be zero. Let the gnomon slide about right and left, up and down, in such a way as never to overlap the right, and the bottom, barrier of the array, although

$$\begin{array}{c|ccccc|}
\cdot & \cdot & \cdot & \cdot & \cdots & \cdot \\
\hline
\cdot & x_{11} & x_{12} & x_{13} & \cdots & x_{1n} \\
\cdot & x_{21} & x_{22} & x_{23} & \cdots & x_{2n} \\
\cdot & x_{31} & x_{32} & x_{33} & \cdots & x_{3n} \\
\vdots & \vdots & \vdots & \vdots & & \vdots \\
\cdot & x_{m1} & x_{m2} & x_{m3} & \cdots & x_{mn} \\
\hline
\end{array}$$

it may overlap the left and top barriers. In this way the gnomon will enclose one, two, or three elements x_{ij}, whose sum is always zero: and its movements give a precise topological account in agreement with equation (20). These facts follow from § 1, p. 143, since the shape of the gnomon and the boundary conditions are determined by the manner in which U' and V enter into the equation (20).

Now consider the mth term T_m of the following sequence:

$$T_1 = [1], \quad T_2 = \begin{bmatrix} . & -1 \\ 1 & -1 \end{bmatrix}, \quad T_3 = \begin{bmatrix} . & . & 1 \\ . & -1 & 2 \\ 1 & -1 & 1 \end{bmatrix},$$

$$T_4 = \begin{bmatrix} . & . & . & -1 \\ . & . & 1 & -3 \\ . & -1 & 2 & -3 \\ 1 & -1 & 1 & -1 \end{bmatrix}, \ldots, \quad . \quad . \quad . \quad . \qquad (21)$$

which evidently satisfies the property $U'X + XU + U'XU = 0$, in the case when $m = n$ and $X = T_m$. It is indeed clear, from the gnomon property of this equation, that the well-known Pascal triangle of binomial coefficients up to the order $(m - 1)$ must be intimately connected with the solution. From this we may readily infer the general solution of (20).*

In fact, by exploring the square array $[x_{ij}]$ in the order $\text{row}_1\rightarrow$, $\text{row}_2\rightarrow$, . . . , it follows that all elements x_{ij} $(i + j < n = m)$ above the secondary diagonal must be zero. Next, whether m and n are equal or not, if X_0 is any solution, so also is aX_0, and again X_0V, and again aX_0V^2, and so on, where a is scaler. Hence the series

$$X = a_0X_0 + a_1X_0V + a_2X_0V^2 + \ldots = X_0f(V), \qquad (22)$$

*Incidentally $T_m{}^3 = I$. Cf. *J. London Math. Soc.* **2** (1927), 242–244.

is a solution, if X_0 is a particular or *key solution*. This is secured by taking

$$\left.\begin{array}{ll} X_0 = T_m & \text{if } m = n, \\ X_0 = \begin{bmatrix} \cdot \\ T_n \end{bmatrix} & \text{if } m > n, \\ X_0 = [\cdot \quad T_m] & \text{if } m < n. \end{array}\right\} \quad . \quad . \quad . \quad . \quad . \quad (23)$$

Solution (22) now resembles (4) in Lemma I, in that it contains an arbitrary row or column—no longer the top row, but, for example, the bottom row in the case when $m = n$.

We infer that (22) is the general solution of (20), and that it contains p arbitrary constants, where p is equal to the smaller of m and n.

The arrays (21) are unsymmetrical: it is, however, possible to base a *symmetrical* solution of (20) upon the matrix

$$S_{2r+1} = \begin{bmatrix} & & & & & & & & \cdot \\ & & & & & & & \iddots & \\ & & & & & & & -2 & \cdots \\ & & & & & & +2 & -5 & \cdots \\ & & & & & -2 & 3 & -4 & \cdots \\ & & & & 2 & -1 & 1 & -1 & \cdots \\ & & & -2 & -1 & \cdot & \cdot & \cdot & \\ & & 2 & 3 & +1 & \cdot & \cdot & \cdot & \\ & -2 & -5 & -4 & -1 & \cdot & \cdot & \cdot & \\ \iddots & \vdots & \vdots & \vdots & \vdots & & & & \\ \cdot & & & & & & & & \end{bmatrix} = S'_{2r+1}, \quad (24)$$

where there are $2r + 1$ rows and columns, with a central $(r + 1, r + 1)$th element 2. Here again the gnomon condition is satisfied, the numerical values of the elements being given by forming the first differences by columns $(\mathrm{col}_{j+1} - \mathrm{col}_j)$ of T_m, and rearranging the results in the manner indicated. (For instance, $[1, 4, 5, 2] = [1, 3, 3, 1] + [\cdot, 1, 2, 1]$.)

The general symmetric solution of (20) is given by

$$X = b_0 X_0 + b_1 U' X_0 U + b_2 U'^2 X_0 U^2 + \ldots + b_r U'^r X_0 U^r = X', \quad (25)$$

where there are $r + 1$ arbitrary constants b_i, and where the key solution is $X_0 = S_{2r+1}$ or $\begin{bmatrix} \cdot & \cdot \\ \cdot & S_{2r+1} \end{bmatrix}$ according as m is odd or even.

The general X is, for example,

$$\underset{(m=3)}{X} = \begin{bmatrix} & & -2b_0 \\ & 2b_0 & -b_0 \\ -2b_0 & -b_0 & 2b_1 \end{bmatrix}, \qquad \underset{(m=4)}{X} = \begin{bmatrix} \cdot & \cdot & \cdot & \cdot \\ \cdot & \cdot & \cdot & -2b_0 \\ \cdot & \cdot & 2b_0 & -b_0 \\ \cdot & -2b_0 & -b_0 & 2b_1 \end{bmatrix}.$$

Returning to the equation (17), let Y be partitioned in accordance with the submatrices of both C and C' which we take to be $\alpha(I+U)$, $\alpha(I+U')$ instead of the usual canonical type $\alpha I + U$, a step which is justified by Ex. 7, p. 155, if $\alpha \neq 0$. This leads us to rewrite equation (17) in the forms

$$\alpha(I+U')Y_{ii}\,\alpha(I+U) = \phi Y_{ii}, \quad Y_{ii} = Y'_{ii}, \quad \alpha \neq 0, \quad . \quad (26)$$

$$\alpha(I+U')Y_{ij}\,\beta(I+V) = \phi Y_{ij}, \quad Y_{ij} = Y'_{ji}, \quad \alpha, \beta \neq 0. \quad . \quad (27)$$

But these equations are of type (18) with a replaced by $(\alpha^2 - \phi)/\phi$ and $(\alpha\beta - \phi)/\phi$ respectively. They therefore admit of the above non-zero solutions (25) and (22), provided that $\phi = \alpha^2$ or $\alpha\beta$ respectively.

If $\alpha = 0$, the submatrix $\alpha I + U$ must take the place of $\alpha(I+U)$: in this case the equations (26) and (27) are slightly modified, but they now can only have a zero solution, seeing that $\phi \neq 0$. In this way the general solution of the original $A'QA = \phi Q$ is ascertained.

EXAMPLES

1. Generating Function of the Key Quadratic Forms.

The quadratic forms whose matrices are $S_1, S_3, S_5, \ldots$ *present themselves as coefficients of powers of* u *in the ascending series for the expansion of*

$$(c_0 + c_1\theta + c_2\theta^2 + c_3\theta^3 + \ldots)(c_0 + c_1u + c_2u^2 + \ldots),$$

where $\theta = -u/(1+u)$, *and* $|u| < 1$, and where all the symbols are scalar.

This function is evidently a quadratic form in the arguments

$$c = \{c_0, c_1, c_2, \ldots\}.$$

Let it first be written as a product of four vector factors

$$c'.\{1, \theta, \theta^2, \ldots\}[1, u, u^2, \ldots].c,$$

which are alternately of the row and column kinds. On combining the middle pair it becomes a double alternant,

$$c'\begin{bmatrix} 1, & u, & u^2, & \ldots \\ \theta, & \theta u, & \theta u^2, & \ldots \\ \theta^2, & \theta^2 u, & \theta^2 u^2, & \ldots \\ \cdot & \cdot & \cdot & \cdot \end{bmatrix} c \equiv c'\Phi c,$$

where Φ can evidently be expanded as an ascending power series in u, with matrix coefficients. Let this be done, and in each term let the scalar u be placed as final factor. We obtain

$$c'\Phi c \equiv$$
$$c'[1]c+c'\begin{bmatrix} & 1 \\ -1 & \end{bmatrix}cu+c'\begin{bmatrix} \cdot & \cdot & 1 \\ 1 & -1 & \cdot \\ 1 & \cdot & \cdot \end{bmatrix}cu^2+c'\begin{bmatrix} \cdot & \cdot & \cdot & 1 \\ -1 & 1 & -1 & \cdot \\ -2 & 1 & \cdot & \cdot \\ -1 & \cdot & \cdot & \cdot \end{bmatrix}cu^3+\ldots,$$

where for shortness the leading and non-zero principal submatrices alone have been written. But, on transposition of this *scalar* expression, we have $c'\Phi'c = c'\Phi c$. By adding together these equal forms we obtain a symmetric coefficient matrix,

$$2c'\Phi c = c'[2]c + c'\begin{bmatrix} \cdot & 1 & 2 \\ 1 & -2 & \cdot \\ 2 & \cdot & \cdot \end{bmatrix}cu^2 + c'\begin{bmatrix} \cdot & -1 & -2 & \cdot \\ -1 & 2 & \cdot & \cdot \\ -2 & \cdot & \cdot & \cdot \\ \cdot & \cdot & \cdot & \cdot \end{bmatrix}cu^3+\ldots$$
$$= s_1 - s_3u^2 + s_3u^3 - s_5u^4 + s_5u^5 - \ldots,$$

where the coefficients s_{2r+1} are evidently the key quadratics in the c_i with the suffixes written in the reverse of the usual order,

$$s_1 = 2c_0^2, \quad s_3 = 2c_1^2 - 2c_0c_1 - 4c_0c_2, \ldots .$$

2. *If* Q *is a symmetric matrix latent in* $\mathrm{C} = \alpha(\mathrm{I}+\mathrm{U})$, $\alpha \neq 0$, *there exists a congruent transformation* $\mathrm{K'QK} = \mathrm{Q_0}$, *such that* K *commutes with* C, *and* $\mathrm{Q_0}$ *is the key solution of*

$$C'QC = \alpha^2 Q.$$

Take
$$K = (c_0I + c_1U + c_2U^2 + \ldots + c_{n-1}U^{n-1}),$$
$$Q = a_0Q_0 + a_1U'Q_0U + \ldots + a_{n-1}U'^{n-1}Q_0U^{n-1},$$

in accordance with the known properties of a commutant (C, C) and a latent Q. Our problem now consists in determining the n coefficients c_i. Write the equation $C'Q_0C = \alpha^2Q_0$ in the alternative form

$$U'Q_0(I+U) = -Q_0U.$$

Since $|I+U| \neq 0$, let $-U/(I+U) = \Theta$. Then $U'Q_0 = Q_0\Theta$, whence $f(U')Q_0 = Q_0f(\Theta)$. The proposed equation, $K'QK = Q_0$, can now be written

$$(c_0I + c_1U' + \ldots)(a_0Q_0 + a_1U'Q_0U + \ldots)(c_0I + c_1U + \ldots) = Q_0.$$

On commuting Q_0 with each power of U' on its left, the corresponding power of Θ is obtained. Thus

$$Q_0(c_0I + c_1\Theta + \ldots)(a_0I + a_1\Theta U + \ldots)(c_0I + c_1U + \ldots) = Q_0,$$

an equation which can be written $Q_0\varphi(U) = Q_0$. On showing that this scalar function φ of the *single* matrix U is in fact unity, the theorem follows by retracing the previous steps.

It therefore remains to choose values of the c_i which reduce the scalar expression,

$$\varphi(u) \equiv (c_0 + c_1\theta + \ldots)(a_0 + a_1\theta u + \ldots)(c_0 + c_1u + \ldots),$$

to unity, where $\theta = -u/(1+u)$. By the previous example

$$\varphi(u) \equiv (s_1 - s_3u^2 + s_3u^3 - s_5u^4 + s_5u^5 - \ldots)(a_0 + a_1\theta u + \ldots).$$

Since $a_0 \neq 0$, the reciprocal $(a_0 + a_1\theta u + \ldots)^{-1}$ may be expanded in ascending powers of u: and, if $\varphi(u) \equiv 1$, this determines the s_i uniquely. From the known s_i successive suitable values of the c_i can be found starting with $c_0 = 1$; whence the matrix K is determined.

The alternate coefficients $c_1, c_3, c_5, \ldots$, may be taken to be zero.

3. *An orthogonal matrix can be found which will transform any orthogonal* A *into a given equivalent orthogonal* B.

Let $C = [\alpha(I + U)]$ be the modified canonical form of both A and B, with $A = HCH^{-1}$, $B = KCK^{-1}$. From the orthogonal properties, $A'A = B'B = I$, it follows that $C'QC = Q$, where $Q = H'H$, and that $C'RC = R$, where $R = K'K$, so that both Q and R are *symmetric*.

Next let a non-singular H_1 be constructed by applying the preceding example to each modified submatrix of C, and combining the results. Then

$$H_1C = CH_1, \qquad H_1'QH_1 = [Q_0],$$

where $[Q_0]$ is the corresponding diagonal assembly of key submatrices. Taking $H_2 = HH_1$, we find that

$$H_2CH_2^{-1} = HH_1CH_1^{-1}H^{-1} = HCH^{-1} = A, \; H_2'H_2 = H_1'H'HH_1 = H_1'QH_1 = [Q_0].$$

Likewise a matrix $K_2 = KK_1$ exists such that

$$K_2CK_2^{-1} = B, \qquad K_2'K_2 = [Q_0].$$

Hence, by eliminating C and $[Q_0]$, both of which are non-singular,

$$PBP^{-1} = A, \qquad P'P = I,$$

where $P = H_2K_2^{-1}$. This is the orthogonal matrix required.

This completes the proof of Ex. 14, p. 109, which only referred to the case of *real* elements.

9. The Resolvent of a Matrix.

The reciprocal of $\lambda I - A$, which evidently exists for all values of λ other than the latent roots of A, is a matrix function of the scalar λ, and is called the *Resolvent* of A,

$$R(\lambda) \equiv \frac{1}{\lambda I - A}. \quad \ldots \ldots \quad (28)$$

Evidently it is a scalar function of the matrix A, and as such is capable of assuming several useful forms. For example, if α is the latent root with maximum modulus,

then $$\frac{1}{\lambda I - A} = \frac{I}{\lambda} + \frac{A}{\lambda^2} + \frac{A^2}{\lambda^3} + \ldots, \quad |\lambda| > |\alpha|. \quad . \quad (29)$$

There is also the interpolation formula of Sylvester, which expresses the function as a polynomial in A, of order $p-1$ or less, and which is based on ψ the R.C.F. of A, where

$$\psi(A) \equiv (A-a_1 I)^{p_{11}}(A-a_2 I)^{p_{12}}\ldots, \qquad \sum_j p_{ij}=p, \quad a_i \neq a_j.$$

EXAMPLES

1. If $\psi(A)=(A-\alpha I)(A-\beta I)^3$, $\alpha\neq\beta$, then

$$\frac{1}{\lambda I-A}=\begin{vmatrix} 1 & 1 & \cdot & \cdot & I \\ \alpha & \beta & 1 & \cdot & A \\ \alpha^2 & \beta^2 & 2\beta & 1 & A^2 \\ \alpha^3 & \beta^3 & 3\beta^2 & 3\beta & A^3 \\ \dfrac{1}{\lambda-\alpha} & \dfrac{1}{\lambda-\beta} & \dfrac{1}{(\lambda-\beta)^2} & \dfrac{1}{(\lambda-\beta)^3} & \cdot \end{vmatrix} \div (\beta-\alpha)^3$$

$$=\frac{A_{11}}{\lambda-\alpha}+\frac{A_{21}}{\lambda-\beta}+\frac{A_{22}}{(\lambda-\beta)^2}+\frac{A_{23}}{(\lambda-\beta)^3}=\frac{B_0\lambda^3+B_1\lambda^2+B_2\lambda+B_3}{(\lambda-\alpha)(\lambda-\beta)^3},$$

where the numerators A_{ij} are cubic polynomials in A, obtained by expanding the determinant by its final row. The B_i are also cubics in A; and the whole process is effected by ordinary scalar algebra.

2. Verify the Hilbert Functional Equation

$$(\lambda-\mu)R(\lambda)R(\mu)=R(\mu)-R(\lambda).$$

We infer that the resolvent $R(\lambda)$ may be regarded in three distinct ways, either as a geometrical progression in A/λ; or as a series of partial fractions $\sum_{i,j}(A_{ij}/(\lambda-a_i)^j)$ summed over the range of values determined by the Segre characteristic of A; or thirdly, as a rational function $B(\lambda)/\psi(\lambda)$, the numerator of which is a polynomial in A and in λ, of order less than p in both. There was implicit allusion to this rational fraction in (22), p. 43, where such a function appeared not as the goal but as the starting-point of an important theorem. Indeed we may adopt what was in fact the original line of development, if we reverse the present argument and deduce the various classical canonical properties from those of the resolvent $R(\lambda)$. Here we shall but sketch the outlines of the theory.

10. The Adjoint Matrix and the Bordered Determinant.

Let x and y be column vectors of order n; then $f=x'Ay$ is a bilinear form in x and y. Now consider the product

$$\begin{bmatrix} I & \cdot \\ x' & 1 \end{bmatrix}\begin{bmatrix} A & \cdot \\ \cdot & -f \end{bmatrix}\begin{bmatrix} I & y \\ \cdot & 1 \end{bmatrix}=\begin{bmatrix} A & Ay \\ x'A & \cdot \end{bmatrix}, \qquad (30)$$

involving matrices of order $n+1$, obtained by attaching the indicated borders of one row and one column to I, or to A. The determinant of the left-hand side product is obviously $-|A|f$. Hence, if $|A| \neq 0$,

$$x'Ay = f = -\begin{vmatrix} A & Ay \\ x'A & . \end{vmatrix} \div |A|.$$

Write $u' = Ay$, $v' = A'x$, so that u and v are row-vectors. Then $vA^{-1}u' = x'Ay$; so that

$$vA^{-1}u' = -\begin{vmatrix} A & u' \\ v & . \end{vmatrix} \div |A| = -\begin{vmatrix} a_{ij} & u'_i \\ v_j & . \end{vmatrix} \div |A|. \quad (31)$$

This is called the *reciprocal* of the bilinear form $x'Ay$. The nature of the numerator determinant may be inferred from Ex. 3, p. 7. Since this identity holds for any non-singular A, we may replace A by $\lambda I - A$ and obtain

$$v(\lambda I - A)^{-1}u' = -\begin{vmatrix} \lambda\delta_{ij} - a_{ij} & u'_i \\ v_j & . \end{vmatrix} \div |\lambda I - A|. \quad (32)$$

The numerator determinant which, upon expansion, is clearly a polynomial of order $n-1$ (or less) in λ, is bilinear in the u_i, v_j, while the denominator is the characteristic function

$$\phi(\lambda) = (\lambda - \alpha)(\lambda - \beta)\ldots(\lambda - \omega) = \psi(\lambda)\,\chi(\lambda), \quad . \quad (33)$$

where $\alpha, \beta, \ldots, \omega$ are the n latent roots, including possible repetitions. When we remove the H.C.F., let us say $\chi(\lambda)$, of numerator and denominator, the relation becomes

$$v(\lambda I - A)^{-1}u' = v\Big(\sum_{i=0}^{p-1} B_i\lambda^{p-i-1}/\psi(\lambda)\Big)u', \quad . \quad (34)$$

where $p(\leqslant n)$ is the order of $\psi(\lambda)$. This gives the rational fractional form $(\lambda I - A)^{-1} = B(\lambda)/\psi(\lambda)$ for the resolvent, which is in fact identical with the result obtained above from the converse point of view, the verification of which we leave to the reader.

EXAMPLES

1. Show (i) that A satisfies the equation $\psi(A) = 0$ as now defined, and no equation of lower order; (ii) that $\psi(A)$ is therefore the R.C.F. of A; (iii) that $\chi(\lambda)$ is identical with the H.C.F. of all $(n-1)$-rowed minors of the determinant $|a_{ij}|$.

[Consider the identity $(\lambda I - A)B(\lambda) = \psi(\lambda)I$. Cf. p. 43.]

2. If $\psi(\lambda)=\lambda^p+c_1\lambda^{p-1}+c_2\lambda^{p-2}+\ldots+c_p$, prove that

$$B_0=I,$$
$$B_1=A+c_1I,$$
$$B_2=A^2+c_1A+c_2I,\ \ldots,$$
$$B_{p-1}=A^{p-1}+c_1A^{p-2}+\ldots+c_{p-1}I.$$

[Use ordinary long division.]

11. Orthogonal Properties of the Partial Resolvents.

The present investigation leads to the classical canonical form through the *orthogonal* properties of the numerators A_{ij} when the function $B(\lambda)/\psi(\lambda)$ is resolved into partial fractions. If all the latent roots are distinct, let us write

$$R(\lambda)=\frac{1}{\lambda I-A}=\frac{R_1}{\lambda-\alpha_1}+\frac{R_2}{\lambda-\alpha_2}+\ldots+\frac{R_n}{\lambda-\alpha_n}=\frac{B(\lambda)}{\phi(\lambda)}, \quad (35)$$

where the n terms are called the *partial resolvents*. They satisfy the orthogonal conditions

$$R_i^2=R_i, \quad R_iR_j=0, \quad i\neq j, \quad . \quad . \quad . \quad . \quad (36)$$

the proof of which depends upon the identities

$$A^r=R_1\alpha_1{}^r+R_2\alpha_2{}^r+\ldots+R_n\alpha_n{}^r, \quad r=0,1,2,3,\ldots \quad . \quad (37)$$

The first n of these relations, which arise by comparing coefficients in descending power series for λ, can be solved by the usual scalar methods. For example

$$R_1=\frac{(A-\alpha_2I)(A-\alpha_3I)\ldots(A-\alpha_nI)}{(\alpha_1-\alpha_2)(\alpha_1-\alpha_3)\ldots(\alpha_1-\alpha_n)}=\frac{g_1(A)}{g_1(\alpha_1)}, \quad (38)$$

let us say, with similar results for all R_i. Hence also

$$(\lambda I-A)B(\lambda)=\phi(\lambda)=(\lambda-\alpha_1)g_1(\lambda)=(\lambda-\alpha_i)g_i(\lambda)$$

identically for all λ. It follows that $\phi(A)=0$, as we already know (p. 41): and also (by the observation that $\lambda-\alpha_i$ must be a factor of the polynomial $g_i(\lambda)-g_i(\alpha_i)$), that $A-\alpha_iI$ is a factor of R_i-I.

Hence $R_i(R_i-I)$ is a polynomial which contains all n factors $(A-\alpha_jI)$, and therefore vanishes with $\phi(A)$; so that $R_i^2=R_i$. Again if $i\neq j$, R_iR_j also contains all these n factors, and therefore vanishes. Thus the orthogonal properties are established.

Analogous, but less elegant, properties may be inferred in the general case of confluent latent roots. The classical canonical form follows by showing that this resolution into partial fractions is virtually equivalent to the now familiar partitioning of the classical $C = HAH^{-1}$ into canonical submatrices. We shall not pursue the matter further, beyond indicating in the following examples the general connexion between the alternative theories.

EXAMPLES

1. If $C = HAH^{-1} = [\alpha_i \delta_{ij}]$, in the simple case (where all α_i differ), then

$$HR_1H^{-1} = \begin{bmatrix} 1 & \cdot & \cdot \\ \cdot & \cdot & \cdot \\ \cdot & \cdot & \cdot \end{bmatrix}, \quad HR_2H^{-1} = \begin{bmatrix} \cdot & \cdot & \cdot \\ \cdot & 1 & \cdot \\ \cdot & \cdot & \cdot \end{bmatrix}, \quad HR_3H^{-1} = \begin{bmatrix} \cdot & \cdot & \cdot \\ \cdot & \cdot & \cdot \\ \cdot & \cdot & 1 \end{bmatrix}:$$

and so on for n rows and columns. Verify the orthogonal properties.

2. Show that $I = R_1 + R_2 + R_3, \quad A = \alpha_1 R_1 + \alpha_2 R_2 + \alpha_3 R_3,$

$$f(A) = f(\alpha_1)R_1 + f(\alpha_2)R_2 + f(\alpha_3)R_3.$$

3. If $HAH^{-1} = C = \begin{bmatrix} \alpha & 1 & \cdot \\ \cdot & \alpha & \cdot \\ \cdot & \cdot & \beta \end{bmatrix}$

and
$$R(\lambda) = \frac{A_{11}}{(\lambda - \alpha)} + \frac{A_{12}}{(\lambda - \alpha)^2} + \frac{A_{21}}{(\lambda - \beta)}, \quad \alpha \neq \beta,$$

then

$$HA_{11}H^{-1} = \begin{bmatrix} 1 & \cdot & \cdot \\ \cdot & 1 & \cdot \\ \cdot & \cdot & \cdot \end{bmatrix}, \quad HA_{12}H^{-1} = \begin{bmatrix} \cdot & 1 & \cdot \\ \cdot & \cdot & \cdot \\ \cdot & \cdot & \cdot \end{bmatrix}, \quad HA_{21}H^{-1} = \begin{bmatrix} \cdot & \cdot & \cdot \\ \cdot & \cdot & \cdot \\ \cdot & \cdot & 1 \end{bmatrix}.$$

4. If in Ex. 3, $\alpha = \beta$, then A_{21} disappears, and $HA_{11}H^{-1} = \begin{bmatrix} 1 & \cdot & \cdot \\ \cdot & 1 & \cdot \\ \cdot & \cdot & 1 \end{bmatrix}$.

If $C = \mathrm{diag}[C_3(\alpha), C_2(\alpha), C_2(\beta)]$, $\alpha \neq \beta$, then

$$HA_{11}H^{-1} = \mathrm{diag}(1, 1, 1;\ 1, 1;\ 0, 0), \quad HA_{12}H^{-1} = \begin{bmatrix} \cdot & 1 & & & & & \\ \cdot & \cdot & 1 & & & & \\ \cdot & \cdot & \cdot & & & & \\ & & & \cdot & 1 & & \\ & & & \cdot & \cdot & & \\ & & & & & \cdot & \cdot \\ & & & & & \cdot & \cdot \end{bmatrix}.$$

$$HA_{21}H^{-1} = \mathrm{diag}(0, 0, 0;\ 0, 0;\ 1, 1),$$

5. In general $A_{ii}^2 = A_{ii}$, $A_{ij} = U_i^{j-i}$, $A_{ij}A_{kl} = 0$, $i \neq k$, where U_i denotes the totality of auxiliary unit submatrices associated with the latent root α_i.

12. Application to Symmetric Matrices. Reduction by Darboux.

Returning to the discussion of the adjoint matrix (p. 162), if $A = \hat{A}'$ we may take $u' = Ay$, $v' = Ax$, and

$$\tilde{x}'Ay = \tilde{v}A^{-1}u' = -\begin{vmatrix} A & u' \\ v & . \end{vmatrix} \div |A| = -\begin{vmatrix} A & Ay \\ \tilde{x}'A & . \end{vmatrix} \div |A|.$$

This leads at once to the reduction, given by Darboux, for expressing a pencil of quadrics as a sum of squares. In the simple case when (i) A_2 is non-singular and (ii) the latent roots are distinct, let

$$A = A_1 + \lambda A_2, \quad |A_2| \neq 0, \quad |A_1 + \lambda A_2| \, |A_2|^{-1} = \prod_{i=1}^{n} (\lambda - \lambda_i) = \phi(\lambda).$$

Then by the ordinary rule for determining the coefficients of partial fractions,

$$-\begin{vmatrix} A & u' \\ v & . \end{vmatrix} \div |A| = -\sum_{i=1}^{n} \begin{vmatrix} A_1 + \lambda_i A_2 & u' \\ v & . \end{vmatrix} \div \{ |A_2| \, (\lambda - \lambda_i)\phi'(\lambda_i) \},$$

since the numerator on the left is of order less than n in λ. Operate on each of these n determinants with

$$\mathrm{col}_{n+1} - \sum_{j=1}^{n} \mathrm{col}_j \,.\, y_j, \quad \mathrm{row}_{n+1} - \sum_{i=1}^{n} \tilde{x}_i' \, \mathrm{row}_i,$$

two processes which leave their values unchanged. We obtain

$$-\sum_i \begin{vmatrix} A_1 + \lambda_i A_2, & (\lambda - \lambda_i) A_2 y \\ (\lambda - \lambda_i)\tilde{x}' A_2, & \theta \end{vmatrix} \div \{ |A_2| \, (\lambda - \lambda_i)\phi'(\lambda_i) \},$$

where θ is scalar. But the cofactor of θ in its determinant is $|A_1 + \lambda_i A_2|$ which is zero. Hence θ may be omitted: and when this is done the factor $(\lambda - \lambda_i)$ may be isolated from the final row, and from the final column, and from the denominator. After cancelling we obtain

$$\tilde{x}'Ay = -\Sigma(\lambda - \lambda_i) \begin{vmatrix} A_1 + \lambda_i A_2, & A_2 y \\ \tilde{x}' A_2, & . \end{vmatrix} \div |A_2| \, \phi'(\lambda_i).$$

But the numerator determinants are now of the type $\begin{vmatrix} B_i & z \\ w & . \end{vmatrix}$, where $|B_i| = 0$: and, by a theorem of Bellavitis, such a type may be resolved into two linear factors $(\Sigma a_i z_i)(\Sigma \beta_j w_j)$. In particular, if B_i is symmetric and $z = w$, these factors are equal. After incorporating the denominators within these equal factors we infer that, when $A = A'$, we may express the pencil $x'Ax$ as $\sum_i (\lambda - \lambda_i)\, \xi_i^2$, where the ξ_i are linear functions (usually complex) of the x_i. Thus

$$x'A_1x = -\sum_i \lambda_i \xi_i^2, \qquad x'A_2x = \sum_i \xi_i^2,$$

which effects the reduction of Darboux, and brings the two quadratics to the sums of squares.

For the confluent case, we may refer the reader to an exposition given by Jessop: *Line Complex* (Cambridge, 1903), 191–8. The full discussion needs the use of bordered determinants of higher orders, and of confluent partial fractions. In the end the result attained is naturally the same as that of Weierstrass (p. 132). The above treatment may be utilized, with the necessary restrictions upon the values of the λ_i, for the case of Hermitian pencils.

13. **Historical Note.**—Matrix equations of the type $AX = XB$ were considered by Sylvester, *Comptes Rendus*, **99** (1884), 67, 115. The type $AX = XA'$ was treated by Cayley, *Mess. Math.*, **14** (1885), 176; and the type $AX = XA$ by Voss, *Sitzb. Akad. Münch.*, **19** (1889), 283–300; and Taber, *Proc. Amer. Ac.*, **26** (1891), 64–6. The general linear equation of many terms was considered by Sylvester, *C. R.*, **99** (1884), 409, 432, 527, 621; and more recently by Maclagan-Wedderburn, *Proc. Edin. Math. Soc.*, **22** (1904), 49, and F. L. Hitchcock, *Proc. Nat. Acad. Sci.*, **8** (1922), 78–83. See further p. 188.

A chapter of Hilton's *Homogeneous Linear Substitutions* (Oxford, 1914) is devoted to linear substitutions permutable with a given substitution.

An account of analytic functions of matrices, as based on the theory of permutable matrices, is given in a paper by H. B. Phillips, *Am. J. Math.*, **41** (1919), 266–78. In this paper a general Taylor series is derived. A thesis, *Om Matrixregning*, by G. Rasch (Copenhagen, 1930), is concerned with the application of matrix functionality to the solution of difference and differential equations; it contains (p. 109–10) a useful bibliography.

P. A. M. Dirac's definition of matrix functions in quantum mechanics by means of permutable matrices is to be found in *Proc. Camb. Phil. Soc.*, **23** (1926), 412–4, or in *Quantum Mechanics* (Oxford, 1930), 41. (The use of U and U' of infinite order on p. 125 is also interesting.)

The use of bordered determinants and partial fractions in the reduction of quadratic forms is first found in a classical memoir of Darboux, *J. de Math.*, **19** (1874), 347–96; the orthogonal properties of the partial resolvents were given by Study, *Monatshefte f. Math. u. Phys.*, **2** (1891), 23–54.

CHAPTER XI

Practical Applications of Canonical Reduction

There is an extensive class of problems in which the behaviour of a function at some critical point depends on the nature of the quadratic terms of small order in the Taylor expansion of the function in the immediate neighbourhood of the point. Many instances of this will occur to mind; for example, the determination of the maxima or minima of a real function of many variables, the discussion of the form of a surface near a given point by means of the indicatrix, the stability of dynamical equilibrium, and so forth. We shall first consider briefly the problem of the maximum or minimum of a quadratic form itself, which we may suppose reduced by a real non-singular congruent transformation to a form involving squares only.

1. The Maximum and Minimum of a Quadratic Form.

The reduced form may be positive definite, non-negative definite of rank r, or nullity $n-r$, negative definite, non-positive definite of rank r, indefinite, or finally null or zero definite. Clearly a positive definite form must have the minimum value zero for zero values of all its variables—at the point zero, as we may say: it is pointwise minimal for any n directions. In the same way a negative definite form is pointwise maximal at zero. A non-negative definite form of rank r is minimal at the value zero in the wide sense that it cannot take negative values. One may say that it is pointwise minimal for certain r directions; but since the r minimal conditions, namely the simultaneous vanishing of its r reduced variables, involve r independent linear homogeneous equations in the n original variables, the solution of which has $n-r$ parameters, we see that the form retains the minimal value zero throughout a certain linear locus of $n-r$ dimensions. A similar remark applies to the maximum of a non-positive definite form of nullity $n-r$. Again, an indefinite form is stationary at zero, but can obviously attain either positive or negative values in the neighbourhood of zero, so that it possesses neither a maximal nor a minimal value. The case of a null or zero definite form is trivial.

These remarks might be amplified by a classification of cases of low order; but a few examples of quadratic forms taken at the point zero will be sufficient illustration.

EXAMPLES

1. The form $a^2x^2 + b^2y^2 + c^2z^2$ is positive definite, pointwise minimal for all three directions at $\{0, 0, 0\}$.

2. The form $(ax + by)^2 + c^2z^2$ is non-negative definite of nullity 1, rank 2, pointwise minimal for two directions, with a minimal line

$$ax + by = 0, \quad z = 0.$$

3. The form $(ax + by + cz)^2$ is non-negative definite of nullity 2, rank 1, pointwise minimal for one direction, also minimal in the plane

$$ax + by + cz = 0.$$

4. The form $a^2x^2 - b^2y^2 + c^2z^2$ is indefinite, neither maximal nor minimal at the point $\{0, 0, 0\}$.

5. Interpret the above, with reference to ellipsoids, cylinders, planes, and hyperboloids, x, y, z being rectangular Cartesian co-ordinates.

If next we add, to a quadratic form, finite cubic or quartic or higher terms in the variables, we shall find that the only functions so formed that remain in general pointwise minimal at zero in all directions are those for which the quadratic terms are positive definite. The same holds in the maximal case for negative definite forms. Consider for example the following functions in the neighbourhoods indicated:

(i) $x^2 + y^2 + z^2 + ax^3 + by^3 + cz^3$, at $\{0, 0, 0\}$ and at $\{\varepsilon, 0, 0\}$, &c.

(ii) $x^2 + y^2 + z^2 + ax^4 + by^4 + cz^4$, at $\{0, 0, 0\}$ and at $\{\varepsilon, 0, 0\}$, &c.

(iii) $(x + y)^2 + z^2 + ax^4$, at $\{0, 0, 0\}$ and at $\{\varepsilon, -\varepsilon, 0\}$.

(iv) $(x + y + z)^2 + ax^3 + by^3$, at $\{0,0,0\}$ and $\{\varepsilon, -\varepsilon, 0\}, \{\varepsilon, 0, -\varepsilon\}$, &c.

The forms in (i) and (ii) are maximal; the form (iii) is maximal if a is positive, and neither maximal nor minimal if a is negative; while the form (iv) is also neither maximal nor minimal.

2. Maxima and Minima of a Real Function.

Consider next a function of several real variables,

$$\phi \equiv \phi(x) \equiv \phi(x_1, x_2 \ldots, x_n), \quad \ldots \quad (1)$$

which possesses finite partial derivatives in the neighbourhood of a point $\xi = \{\xi_i\}$ at which the first derivatives vanish together, so that we have, at $x = \xi$,

$$\frac{\partial \phi}{\partial x_i} = 0, \qquad i = 1, 2, \ldots, n. \quad . \quad . \quad . \quad . \quad (2)$$

Let us adopt the notation, always at the point $x = \xi$,

$$\phi_i = \frac{\partial \phi}{\partial x_i}, \quad \phi_{ij} = \frac{\partial^2 \phi}{\partial x_i \partial x_j} = \phi_{ji}, \qquad h = \{h_1, h_2, \ldots, h_n\},$$

$$f = \{\phi_1, \phi_2, \ldots, \phi_n\}, \quad F = [\phi_{ij}] = F', \quad . \quad . \quad (3)$$

where h represents a vector of increments h_i, all of the same order of smallness. The Taylor series for ϕ at $\xi + h$ may then be written

$$\phi(\xi + h) - \phi(\xi) = h'f + \frac{1}{2!}h'Fh + R, \quad . \quad . \quad . \quad (4)$$

where R is a remainder consisting of terms of the third and higher orders.

A necessary condition for a maximum or minimum of ϕ at ξ is that the vector f should be null; for this ensures that ϕ is stationary at ξ. But whether there is a maximum or a minimum depends first of all on the quadratic terms $h'Fh$ for those values of f_{ij} for which $f = 0$. By what has preceded we may assert:

(i) *If* F *is positive definite, ϕ is a minimum.*

(ii) *If* F *is negative definite, ϕ is a maximum.*

(iii) *If* F *is indefinite, ϕ is neither maximum nor minimum.*

In the excluded cases the nature of ϕ at the critical point ξ cannot in general be ascertained without the consideration of the terms in R, a course which must in any case be adopted if F is null. The limited statement might be made that, if F is non-negative definite of rank r, then ϕ is minimal in r co-ordinates, which are linear functions of the variables x_i, and that it is so far undetermined in the remaining $n - r$, with a corresponding remark in the non-positive definite case.

EXAMPLES

1. Consider the maxima and minima of Hermitian forms.

2. Examine the maxima and minima of the functions:

   (i) $x^2y^3(y^2 + y - 1)$, (ii) $x^3 + y^3 - 3xy$, (iii) $x^2 + xy + y^2 - ax - by$, (iv) $(x^2 - 2ax)(y^2 - 2by)$.

3. Find the maximum value of $(ax + by + cz)e^{-\frac{1}{2}(x^2+y^2+z^2)}$.

3. Conditioned Maxima and Minima of Quadratic Forms.

In many applications it is required to find the maximum or minimum of a quadratic or Hermitian form $x'Ax$, subject to one or more restrictive conditions. Consider for example the problem of determining the lengths of the major and the minor semi-axes of an ellipse which has the equation in rectangular co-ordinates

$$ax^2 + 2hxy + by^2 = 1, \quad ab > h^2. \quad . \quad . \quad . \quad (5)$$

Let us denote the radius vector by r, the vector $\{x, y\}$ by x simply, the matrix of the quadratic form by A, and the vector of direction-cosines $\{\cos\theta, \sin\theta\}$ by p. The problem is then to determine the maximum and minimum of

$$p'Ap = 1/r^2, \quad . \quad . \quad . \quad . \quad . \quad . \quad . \quad . \quad (6)$$

under the condition

$$p'p = 1. \quad . \quad . \quad . \quad . \quad . \quad . \quad . \quad . \quad . \quad (7)$$

Introducing a Lagrange multiplier λ, we are led to consider the quadratic form

$$p'Ap - \lambda p'p, \quad . \quad . \quad . \quad . \quad . \quad . \quad . \quad . \quad (8)$$

of matrix $A - \lambda I$. The condition for maximum or minimum is then

$$(A - \lambda I)p = 0, \quad . \quad . \quad . \quad . \quad . \quad . \quad . \quad . \quad (9)$$

a set of equations which has for its solution a non-zero vector p when, and only when, λ is a latent root of A. We have then, for such a vector $p_{(\lambda)}$,

$$p'_{(\lambda)}Ap_{(\lambda)} = \lambda p'_{(\lambda)}p_{(\lambda)} = \lambda, \quad . \quad . \quad . \quad . \quad (10)$$

which shows that the maximum and minimum values of r are $\lambda_1^{-\frac{1}{2}}$ and $\lambda_2^{-\frac{1}{2}}$, where λ_1 and λ_2 are the latent roots of A. These values for the semi-axes are of course real, the quadratic form $x'Ax$ being positive definite. The extension to the n-dimensional case is evident.

The problem of the maximum and minimum of an Hermitian form $\bar{x}'Ax$ under the condition $\bar{x}'x = 1$ is precisely analogous, the required values being the greatest and the least latent roots of A, namely λ_n and λ_1. Reciprocally it may be seen that the maximum and the minimum of $\bar{x}'x$ subject to the condition $\bar{x}'Ax = 1$ are the greatest and the least roots of the equation $|\mu A - I| = 0$.

A more general problem involves the maximum or minimum of a form $\bar{x}'Ax$, subject to the condition $\bar{x}'Bx = 1$, where one or other of the matrices A and B is positive or negative definite. It is clear that in this case the matrix $A - \lambda I$ of the earlier discussion will be replaced by $A - \lambda B$, the critical values λ being roots of $|A - \lambda B| = 0$.

For example, find the maximal and minimal values of the radius vector of an ellipse which, referred to oblique axes inclined at an angle ω, has the equation

$$ax^2 + 2hxy + by^2 = 1, \qquad ab > h^2.$$

The solutions of the minimal problem which involve latent roots, other than the greatest and the least, refer to successively added conditions. It may be shown, for example, by introducing further Lagrange multipliers, that the mth least root, λ_m, is the minimal value of the form $\bar{x}'Ax$ subject to the set of unitary conditions

$$\bar{x}'x = 1, \quad \bar{x}'_{(\lambda_i)}x = 0, \qquad i = 1, 2, \ldots, m-1, \quad . \quad (11)$$

with a corresponding extension to the case of the condition $\bar{x}'Bx = 1$ and other added conditions, where B is positive definite. The problem may be realized by considering the principal semi-axes of a general n-dimensional ellipsoid, in rectangular or oblique co-ordinates.

4. The Vibration of a Dynamical System about Equilibrium.

Closely related to the preceding is the problem of determining the motion of a dynamical system of a finite number n of degrees of freedom when displaced slightly from rest. Denoting by generalized co-ordinates q_i the displacement from a position of equilibrium, at which $q_i = 0$, $\dot{q}_i = 0$, $\ddot{q}_i = 0$, we know (cf. Whittaker, *Analytical Dynamics*, p. 178) that the kinetic energy T for a small displacement $\{q_i\}$ is a homogeneous quadratic in the $\dot{q}_i$, which from the very nature of kinetic energy must be positive definite: while the potential energy V with respect to the position of equilibrium is a quadratic in q, not necessarily positive definite. Further the $\dot{q}_i$, being derivatives of the q_i, are cogrediently transformed with them. It is assumed that T and V do not involve the time explicitly.

The problem then is the simultaneous reduction of the pair of forms

$$V = q'Aq, \qquad T = \dot{q}'B\dot{q}, \quad . \quad . \quad . \quad . \quad . \quad (12)$$

where B is positive definite. By p. 107, Chapter VIII, these may be transformed together by a real congruent transformation into

$$V = \sum_{i=1}^{n} \alpha_i \eta_i^2, \qquad T = \sum_{i=1}^{n} \dot{\eta}_i^2. \quad . \quad . \quad . \quad . \quad . \quad (13)$$

In these new normal co-ordinates the Lagrangian equations of motion

$$\frac{d}{dt}\left(\frac{\partial T}{\partial \dot{q}_i}\right) - \frac{\partial T}{\partial q_i} = -\frac{\partial V}{\partial q_i} \quad . \quad . \quad . \quad . \quad . \quad (14)$$

take the shape

$$\frac{d^2\eta_i}{dt^2} + a_i\eta_i = 0, \qquad i = 1, 2, \ldots, n. \quad . \quad . \quad . \quad (15)$$

Now the roots a_i are all real, since B is positive definite (Ex. 6, p. 108). If further they are all positive, then the system has n superposed small harmonic vibrations about equilibrium,

$$\eta_i = c_i \cos \sqrt{a_i}(t - \epsilon_i). \quad . \quad . \quad . \quad . \quad . \quad (16)$$

In this case the equilibrium is stable; and evidently (Ex. 7, p. 108) the condition for this is, as one might expect, that both matrices A and B should be positive definite, that is, that V should be a positive definite form as well as T.

If A is merely non-negative definite of rank $r < n$, then in general the equilibrium is unstable, for (corresponding to certain necessarily zero values of a_i) in certain of the new co-ordinates there will be motions of the type

$$\eta_i = c_i t + \beta_i; \quad . \quad . \quad . \quad . \quad . \quad . \quad (17)$$

while if A is indefinite or non-positive or negative definite, so that certain of the roots a_i are negative, there will be motions

$$\eta_i = c_i \cosh \sqrt{-a_i}(t - \epsilon_i), \quad . \quad . \quad . \quad . \quad . \quad (18)$$

indicating a still more marked type of instability.

EXAMPLES

1. A rod of length $2a$ is suspended from a fixed point by a string of length h attached to its end. Prove that the rod, when slightly disturbed, has two normal vibrations; and find their periods.

[If θ and φ are the angles which the string and the rod make with the vertical when displaced, then

$$V = \tfrac{1}{2}mg(h\theta^2 + a\varphi^2), \quad T = \tfrac{1}{2}m(h^2\dot{\theta}^2 + 2ha\dot{\theta}\dot{\varphi} + a^2\dot{\varphi}^2 + k^2\dot{\varphi}^2),$$

where $mk^2 = \frac{1}{3}ma^2$ is the moment of inertia of the rod about its middle point.]

2. A system which would otherwise vibrate about stable equilibrium in n normal modes of periods $2\pi/\lambda_i$, where $\lambda_1 < \lambda_2 < \ldots < \lambda_n$, receives a constraint which reduces the degrees of freedom to $n - 1$. Prove that if the new periods are $2\pi/\mu_i$, the $n - 1$ values of μ alternate with the n values of λ.

[By a suitable choice of co-ordinates the equation for the constraint may be written $q_n = 0$.]

5. Matrices and Quadratic Forms in Mathematical Statistics.

The mathematical apparatus of statistics has much in common with the dynamics of systems of particles and of rigid bodies; for example the arithmetic mean, fundamental in statistics, is simply the centroid, the standard deviation is the radius of gyration, moments and product moments appear in both, and so on. Many of the most convenient parameters for descriptive purposes in statistics involve quadratics, and the very important normal function of frequency is the exponential of a negative definite quadratic form. Again, the principle of Least Squares implies in its name the minimizing of a positive definite quadratic form. We shall show by a few elementary examples that the notation of vectors, matrices, and quadratic forms is well adapted for use in this department.

We may first note the obvious fact that the vector of partial derivatives with respect to the variables u_i of a bilinear form uAx is simply Ax, while that of the partial derivatives with respect to the x_i is uA. In fact the rule for differentiating with respect to vectors is formally the same as the ordinary scalar rule. Again, for a quadratic form $x'Ax$, the vector of partial derivatives with respect to the x_i is $2x'A$, or in column form $2Ax$.

The Principle of Least Squares

In the theory of Least Squares we have a system of observational equations

$$Ax = h, \quad . \quad . \quad . \quad . \quad . \quad . \quad . \quad . \quad . \quad (19)$$

in which there are more equations than unknowns x. The equations are also inconsistent, because of accidental errors of observation. If we rectify this by writing

$$Ax - h = \epsilon, \quad . \quad . \quad . \quad . \quad . \quad . \quad . \quad . \quad (20)$$

where ϵ is a vector of errors ϵ_i, then the principle of Least Squares postulates that *the most satisfactory values of the* x_i *are those for which the sum of the squares of the errors is a minimum.* But this sum is simply $\epsilon'\epsilon$, namely

$$\epsilon'\epsilon = (x'A' - h')(Ax - h). \quad . \quad . \quad . \quad . \quad (21)$$

The minimal conditions, obtained by partially differentiating this with respect to the vector x and halving, give the *normal equations*

$$A'Ax = A'h, \quad . \quad . \quad . \quad . \quad . \quad . \quad . \quad . \quad (22)$$

the relation of which to the original observational equations is at once apparent.

In the above it has been assumed that the errors ϵ_i are of equal importance and are uncorrelated. If this were not so we should have to minimize not the sum of squares $\epsilon'\epsilon$ but some positive definite quadratic form $\epsilon'B\epsilon$, that is,

$$\epsilon'B\epsilon = (x'A' - h')B(Ax - h). \quad . \quad . \quad . \quad . \quad (23)$$

The minimal equations or *normal equations of correlated Least Squares* are therefore

$$A'BAx = A'Bh. \quad . \quad . \quad . \quad . \quad . \quad . \quad . \quad (24)$$

The matrices $A'A$, or more generally $A'BA$, of the normal equations are symmetric and, in all practical cases, positive definite.

Quadratic Moments and Total Correlation Coefficients

The matrix notation lends itself also to the expression of *quadratic moments*, which play so important a part in the theory of normal linear correlation. Suppose that a large number m of samples, taken from a normal homogeneous stock, are measured in respect of n correlated characters. The results might be recorded in a table of m row-vectors each of n real elements, where $m > n$. If we regard this as a matrix X of order $m \times n$, then the symmetric matrix $X'X$ has sums of squares in the diagonal, and sums of products of sample measures everywhere else. It is the matrix of quadratic moments. By means of a diagonal matrix D we may normalize it by a congruent transformation so that the diagonal elements become units; thus

$$D'X'XD = R, \quad R = R', \quad r_{ii} = 1. \quad . \quad . \quad . \quad (25)$$

This matrix R is the *matrix of total correlation coefficients* of the n variables x_i as computed from sample measures. It is seen at once (Ex. 2, p. 97) to be positive definite, and such that each principal minor determinant cannot exceed any principal minor contained in it; in particular $|R| \leqslant 1$. The theory of partial correlation is based on the minors of this matrix.

The Normal Frequency Function

Again, the normal frequency function of many correlated variables can be derived in the form

$$\phi(x) = c \exp\left(-\tfrac{1}{2}x'Ax\right), \quad . \quad . \quad . \quad . \quad . \quad (26)$$

where A is symmetric and positive definite. We fix the constant c by observing that the multiple integral of the function over the *vector range* $\{-\infty\}$ to $\{\infty\}$ is unity. The integral is easily evaluated. By a real congruent transformation $H'AH$ we may reduce the form $x'Ax$ to a sum of squares; thus

$$x = H\xi, \quad x'Ax = \xi' H'AH\xi = \xi'\xi, \quad H'AH = I. \tag{27}$$

The Jacobian of the transformation is then

$$\left|\frac{\partial(x)}{\partial(\xi)}\right| = |H| = |A|^{-\frac{1}{2}}, \text{ since } H'AH = I. \tag{28}$$

The integral thus becomes a product of n separate integrals, thus:

$$\int_{\{-\infty\}}^{\{\infty\}} \exp(-\tfrac{1}{2}x'Ax)\,dx = |A|^{-\frac{1}{2}} \int_{\{-\infty\}}^{\{\infty\}} \exp(-\tfrac{1}{2}\xi'\xi)\,d\xi$$

$$= |A|^{-\frac{1}{2}} \prod_{i=1}^{n} \int_{-\infty}^{\infty} \exp(-\tfrac{1}{2}\xi_i^2)\,d\xi_i$$

$$= (2\pi)^{\frac{1}{2}n}|A|^{-\frac{1}{2}}, \tag{29}$$

by the known result for a single variable. In fact the function of multiple correlated frequency, in which the variables are normalized in order to have unit standard deviation, is

$$\phi(x) = (2\pi)^{-\frac{1}{2}n}|A|^{\frac{1}{2}} \exp(-\tfrac{1}{2}x'Ax). \tag{30}$$

Mean Values of Sample Quadratic Moments

The quadratic moments are quadratic forms in the sample variables. It is important in statistics to know what is the mean value of a parameter, regarded as arising from all possible samples of N measurements. Suppose that in a multiple normal distribution any quadratic parameter $x'Bx$ is under consideration, where the vector x has Nn components, ordered in groups of n. The mean value of the parameter will then be an Nn-ple integral, of the form

$$(2\pi)^{-\frac{1}{2}Nn}|A|^{\frac{1}{2}} \int_{\{-\infty\}}^{\{\infty\}} x'Bx \exp(-\tfrac{1}{2}x'Ax)\,dx. \tag{31}$$

But this is the coefficient of λ in the power series for the integral

$$(2\pi)^{-\frac{1}{2}Nn}\,|\,A\,|^{\frac{1}{2}}\int_{\{-\infty\}}^{\{\infty\}}\exp\,(-\tfrac{1}{2}x'Ax+\lambda x'Bx)\,dx, \quad . \quad (32)$$

where λ may be given such a value, arbitrarily small if necessary, that the matrix $A - 2\lambda B$ is still positive definite. Again the value of this integral, as has just been proved, is the determinant

$$|\,A\,|^{\frac{1}{2}}\,|\,A-2\lambda B\,|^{-\frac{1}{2}}. \quad . \quad . \quad . \quad . \quad . \quad . \quad . \quad (33)$$

Now, neglecting terms of higher than the first degree in λ, we have

$$(I-2\lambda A^{-1}B)^{-\frac{1}{2}} = I+\lambda A^{-1}B+\ldots, \quad . \quad . \quad (34)$$

so that the value of the integral is the coefficient of λ in $|\,I+\lambda A^{-1}B\,|$ Thus the mean value in question is the sum of the diagonal elements in the matrix $A^{-1}B$, or, equally well, in BA^{-1}.

The above elementary instances go to show that the methods of matrices and quadratic forms can be applied with advantage to problems of statistical frequency.

EXAMPLES

1. The value of the integral

$$(2\pi)^{-1}\int\!\!\int_{-\infty}^{\infty}(ax^2+2hxy+by^2)e^{-\frac{1}{2}(x^2+y^2)}\,dx\,dy$$

is $a+b$, as is otherwise obvious.

2. The *variance*, or squared standard deviation, about the mean of N sample values x_i of a variable x is $\frac{1}{N}\Sigma x_i^2 - \frac{1}{N^2}(\Sigma x_i)^2$. If the frequency function is

$$\varphi(x) = \sigma^{-1}(2\pi)^{-\frac{1}{2}}e^{-\frac{1}{2}x^2/\sigma^2},$$

prove that the mean value of the sample variance is $\frac{N-1}{N}\sigma$.

6. Sets of Linear Operational Equations with Constant Coefficients.

A consistent set of linear differential or difference equations in several unknown functions of a variable, with constant coefficients, can be solved by a procedure exactly analogous to Smith's reduction for λ-matrices. If the operation of differentiating with respect to x, or of differencing at equidistant intervals in any of its forms, be denoted by θ, the problem is to find the n functions $y_i(x)$ from the n equations,

$$a_{i1}(\theta)\, y_1 + a_{i2}(\theta)\, y_2 + \ldots + a_{in}(\theta)\, y_n = p_i(x), \quad i = 1, 2, \ldots, n, \quad (35)$$

where the operators a_{ij} are polynomials in θ with constant coefficients. Let the sets of functions y_i and p_i be written as column vectors y and p, and let the θ-matrix of operators be denoted by A. Then the equations take the form

$$Ay = p. \quad . \quad . \quad . \quad . \quad . \quad . \quad . \quad . \quad (36)$$

Since the operator θ, when combined with constant coefficients in a polynomial, is subject to the ordinary rules of algebra, the matrix A has the properties of a λ-matrix. It may therefore be reduced as in Chapter III, p. 23, to the diagonal matrix

$$HAK = [E_i(\theta)\,\delta_{ij}] = F, \quad |H| = 1, \quad |K| = 1, \quad . \quad (37)$$

H and K being θ-matrices, where the reduction takes place in the field of the coefficients of the equations, and the diagonal polynomials $E_i(\theta)$ are the invariant factors of A. Putting $K^{-1}y = z$, $Hp = q$, we have therefore

$$Fz = q. \quad . \quad . \quad . \quad . \quad . \quad . \quad . \quad . \quad (38)$$

But this is a set of equations in which the variables z_i are isolated, and each equation can be solved independently of the rest by the regular methods. Finally the required solutions y_i are given by

$$y = Kz. \quad . \quad . \quad . \quad . \quad . \quad . \quad . \quad . \quad (39)$$

The consistency of a rectangular system of m equations in n unknowns may also be treated in the manner of p. 29.

The number of arbitrary constants entering into a solution z_i is equal to the degree of $E_i(\theta)$, so that the total number of such constants in the complete solution is the degree in θ of the determinant $|F|$, or $|A|$. Again, since the determinant $|K|$ is independent of θ, at least one of the elements in each row of the matrix K must contain a non-zero constant, so that in the operation $y = Kz$ no arbitrary constant in z is entirely obliterated by powers of θ. Hence the number of arbitrary constants in the solution y is again equal to the degree in θ of $|A|$.

The reduction to a diagonal system Fz is not utterly necessary. A semi-reduction to triangular shape $HA = \Gamma$, where $\gamma_{ij} = 0$, $i > j$, is sufficient for successive solution, beginning with the last equation.

EXAMPLES

1. Solve the simultaneous differential equations:

$$\begin{aligned} Dy_1 - Dy_2 &= e^x, \\ 2D^2 y_1 - D(D-1)y_2 - D(D+1)y_3 &= 3e^x \\ 3Dy_1 + 3D^2y_2 + D(D^2-1)y_3 &= 7e^x, \end{aligned}$$

where $D = \dfrac{d}{dx}$.

2. Solve the simultaneous difference equations:

$$\begin{aligned} \Delta y_1 + 2\Delta^2 y_2 + 3\Delta y_3 &= 3^x, \\ \Delta y_1 + \Delta(\Delta - 1)y_2 - 3\Delta^2 y_3 &= 3^x - 2^x, \\ \Delta(\Delta+1)y_2 - \Delta(\Delta^2-1)y_3 &= 1, \end{aligned}$$

where $\Delta y(x) = y(x+1) - y(x)$.

7. Historical Note.—Recent accounts of the problem of maxima and minima of quadratic forms under conditions are to be found in a paper of R. G. D. Richardson, *Trans. Am. Math. Soc.*, **26** (1924), 479–94, and Courant and Hilbert, *Mathematische Physik I* (Berlin, 1924), 228–31, the latter treating of vibrations and resonance of dynamical systems.

The reduction of the pair of forms in the problem of small vibrations and the demonstration of the linear nature of the elementary divisors are due to Weierstrass: *Berlin. Monatsb.* (1858), 207, or *Werke*, I, 233.

An ample discussion of the application of matrices and quadratic forms to problems of statistical correlation may be found in a paper by R. Frisch, *Nordisk Statist. Tidskr.*, **8** (1928), 36–102. A vector notation is used, but row and column vectors are not distinguished.

The solution of a set of simultaneous linear differential equations with constant coefficients was obtained by Chrystal, *Trans. Roy. Soc. Edin.*, **38** (1895), 163, by the semi-reduction referred to above. For a good account of this and fuller references the textbook of E. L. Ince, *Ordinary Differential Equations* (London, 1927), may be consulted.

The matrix operator of Cayley (*Invariants*, p. 113 and p. 72), which has partial differential operators for elements, has been studied in relation to matrix operands and invariants by H. W. Turnbull, *Proc. Edin. Math. Soc.* (2), **1** (1927), 111–28, **2** (1929), 33–54, 256–64, *Proc. London Math. Soc.* (2), **33** (1931), 1–21. It has also been used by J. Williamson to deduce new results concerning the determinants of Bazin and Reiss; *Proc. Edin. Math. Soc.* (2), **2** (1929), 240–51.

Matrices of infinite order, which we do not consider in the present book, came to be studied when integral equations were first investigated, at the beginning of the present century. The subject is of vast range and is of fundamental importance. Great impetus was given to it by an application in 1925 to quantum mechanics, in the hands of Heisenberg, Born, and Jordan.

A bibliography of writings on matrices up to 1928 has been compiled by Muir, *Trans. Roy. Soc. S. Africa*, **18** (1929), 219–27.

Of more recent textbooks on matrices, we may cite *The Theory of Matrices* (Berlin, 1933) by C. C. MacDuffee, and *Lectures on Matrices* (*Amer. Math. Soc.*, Colloquium Publications, 1934) by J. H. M. Wedderburn, which contains an extensive bibliography.

APPENDIX

Further Rational Canonical Forms of a Matrix

The methods employed in Chapter V and also in finding a real classical canonical form for a real matrix (p. 72) can readily be extended to the case when the reduced characteristic function $\psi(\lambda)$ of a matrix A has been resolved as far as possible into its irreducible factors within a prescribed field. Suppose, for example, that ψ has three distinct irreducible factors ψ_1, ψ_2, ψ_3 within the real rational field $\mathcal{F}$, so that no pair of these have any further factor in common. Let their degrees be p_1, p_2, p_3 respectively, so that $p_1 + p_2 + p_3 = p$, the degree of ψ. In this case the leading $p \times p$ submatrix B_p ((11), p. 49) of the rational canonical form can be broken down into three further isolated submatrices $B_{p_1}, B_{p_2}, B_{p_3}$ within the field $\mathcal{F}$.

To establish this result let column vectors x, y, z be employed, as in Ex. 6, p. 57. Henceforward let $\psi, \psi_1, \psi_2, \psi_3$ denote the matrix polynomials $\psi(A)$, etc., so that each is a matrix of order $n \times n$ that of A, and all such matrices commute. Since $\psi_1\psi_2\psi_3 = \psi$, the R.C.F. of A, the product vanishes: moreover, each ψ_i is a singular matrix else $\psi_i^{-1}\psi$ would be a vanishing polynomial in A of degree less than p, which cannot be. Hence non-zero vectors x, y, z exist such that

$$\psi_1(A)x = 0, \quad \psi_2(A)y = 0, \quad \psi_3(A)z = 0. \qquad (1)$$

By Theorem I, p. 47, ψ_1 must therefore contain the R.C.F. X of x as a factor; but ψ_1 has no factors. Hence ψ_1 is the R.C.F. of x. Similarly for y and for z. From each of these vectors a chain can be formed in the usual way, such as x, Ax, A^2x, etc., consisting respectively of p_i vectors which are linearly independent in $\mathcal{F}$, each chain terminating because of a condition (1). Let a matrix be obtained by reversing these chains and then combining them into an extended chain of length p:

$$H_p = [A^{p_1-1}x, \ldots, Ax, x, A^{p_2-1}y, \ldots, Ay, y, A^{p_3-1}z, \ldots Az, z] \quad (2)$$

so that H_p has $p_1 + p_2 + p_3 = p$ columns and n rows. Then these p columns of H_p are linearly independent in $\mathcal{F}$. For if not, a linear relation between them must exist which, in view of the way that x, y, z enter into (2), can be written

$$Lx + My + Nz = 0, \qquad (3)$$

where L, M, N are polynomials in A of degrees less than p_1, p_2, p_3 respectively, since A never reaches these degrees in the expression (2). Whereas one or more but not all three of L, M, N may be zero, let $L \neq 0$. Premultiply (3) by $\psi_2\psi_3$, remembering that these matrices commute with L, M, and N. Then by (1) $L\psi_2\psi_3 x = 0$ and hence the matrix $L\psi_2\psi_3$ by annihilating x must contain ψ_1 as a factor (Theorem I, p. 47): that is, say, $L\psi_2\psi_3 \equiv P\psi_1$. Since the degree of L is less than p_1, that of the polynomial P must be less than $p_2 + p_3$, as we see by comparing degrees on both sides of this identity. Also P must contain $\psi_2\psi_3$ as a factor, which is impossible since it is of degree lower than that of $\psi_2\psi_3$. Thus no such identity (3) can exist.

On premultiplying H_p by A we can at once establish the requisite canonical form by means of the identity

$$AH_p = H_p \operatorname{diag}(B_{p_1}, B_{p_2}, B_{p_3}), \quad . \quad . \quad . \quad . \quad (4)$$

which is exemplified by carrying out the details as follows:

Suppose

$$\psi = \psi_1\psi_2\psi_3 = (A^2 - c_1A - c_2I)(A^3 - d_1A^2 - d_2A - d_3I)(A^2 - e_1A - e_2I),$$

where $p_1 = 2$, $p_2 = 3$, $p_3 = 2$, and $p = 7$, then

$$AH_7 = [Ax,\ x,\ A^2y,\ Ay,\ y,\ Az,\ z]\left[\begin{array}{cc|ccc|cc} c_1 & 1 & \cdot & \cdot & \cdot & \cdot & \cdot \\ c_2 & \cdot & \cdot & \cdot & \cdot & \cdot & \cdot \\ \hline \cdot & \cdot & d_1 & 1 & \cdot & \cdot & \cdot \\ \cdot & \cdot & d_2 & \cdot & 1 & \cdot & \cdot \\ \cdot & \cdot & d_3 & \cdot & \cdot & \cdot & \cdot \\ \hline \cdot & \cdot & \cdot & \cdot & \cdot & e_1 & 1 \\ \cdot & \cdot & \cdot & \cdot & \cdot & e_2 & \cdot \end{array}\right] \quad . \quad . \quad (5)$$

Here $n \geqslant 7$.

The comparison of both sides of (4), column by column, immediately verifies this identity, except for the leading column of each submatrix. But these on comparison are true owing to the relations (1); for example, the first column of AH_7 is A^2x, which is $(c_1A + c_2)x$, which in turn yields the first column on the right of (5) (cf. Ex. 5, p. 57).

Manifestly this procedure is applicable to any number, two or more, of factors ψ_i belonging to the R.C.F. ψ, provided that no pair of them has a common factor. Indeed, with this provision it is not necessary for each such ψ_i to be irreducible. It is left to the reader to examine what extra safeguard is then necessary in choosing x, y, z to satisfy (1): cf. Ex. 1, p. 56.

The Case of Repeated Irreducible Factors.

To complete the discussion we must consider the case when ψ has repeated factors which are irreducible in a field $\mathcal{F}$. Since we may isolate the parts of the canonical matrix that belong to the distinct factors of ψ, it is enough to consider the case of a single factor in repetition; for example $\psi = \chi^3$, where χ is a polynomial of degree q in A and ψ is of degree p, so that $p = 3q$.

Since ψ is the R.C.F. of A there must exist a vector x of maximal grade p (Theorem IV, p. 52), which is the degree of ψ, such that no polynomial $L = L(A)$ in A of degree less than p can annihilate x. Now consider the chain

$$H = [A^{q-1}z, \ldots, Az, z, A^{q-1}y, \ldots, Ay, y, A^{q-1}x, \ldots, Ax, x],$$

where $z = \chi y$ and $y = \chi x$. Expressed in terms of x this is

$$[A^{q-1}\chi^2 x, \ldots, A\chi^2 x, \chi^2 x, A^{q-1}\chi x, \ldots, A\chi x, x, A^{q-1}x, \ldots, Ax, x],$$

where χ^2 occurs in the first q consecutive components, χ in the next q, but is absent in the final q components. Furthermore, the cofactors of x are polynomials or single terms in A, all of positive or zero degree less than p, the highest being of degree $3q - 1 = p - 1$. Hence, as before, these vectors, p in number, which form the chain are linearly independent.

Again, we obtain a canonical form, analogous to (5), only in this case the units that appear upon the superdiagonal present a continuous run $p - 1$ in number. This is sufficiently exemplified as follows:

Take $\qquad \psi(A) = (A^3 - d_1A^2 - d_2A - d_3I)^3 = \chi^3$

and a maximal vector x of grade 9, so that $\chi^3 x = 0$. Thereupon take $y = \chi x$ and $z = \chi^2 x$. Then $AH = HB$, where

$$H = [A^2z, Az, z, A^2y, Ay, y, A^2x, Ax, x]$$

$$B = \left[\begin{array}{ccc|ccc|ccc} d_1 & 1 & . & . & . & . & . & . & . \\ d_2 & . & 1 & . & . & . & . & . & . \\ d_3 & . & . & 1 & . & . & . & . & . \\ \hline . & . & . & d_1 & 1 & . & . & . & . \\ . & . & . & d_2 & . & 1 & . & . & . \\ . & . & . & d_3 & . & . & 1 & . & . \\ \hline . & . & . & . & . & . & d_1 & 1 & . \\ . & . & . & . & . & . & d_2 & . & 1 \\ . & . & . & . & . & . & d_3 & . & . \end{array}\right] \qquad \ldots \quad (7)$$

Again, all but the first, fourth, and seventh columns are immediately verifiable in the left and right sides of the identity $AH = HB$, while these cited columns give A^3z, A^3y, A^3x on the left, and

$$(d_1A^2 + d_2A + d_3I)z$$
$$z + (d_1A^2 + d_2A + d_3I)y$$
$$y + (d_1A^2 + d_2A + d_3I)x$$

on the right respectively: which agree, since $\chi z = 0$, $z = \chi y$, and $y = \chi x$.

Rational Commutants.

The solution of the equation $AX = XA$ of p. 146 was irrational, because it was given in terms of the latent roots of the $n \times n$ matrix A, whereas the n^2 equations for the elements x_{ij} of X, of which $\sum_{k=1}^{n} a_{ik}x_{kj} = \sum_{k=1}^{n} x_{ik}a_{kj}$ is typical, are linear and rational. To obtain a rational solution let $HAH^{-1} = B$ be a rational canonical form of A, and let $Y = HXH^{-1}$ so that $BY = YB$. On solving this rationally we can then obtain a rational X in the form $H^{-1}YH$.

Now let $B = \text{diag}(B_p, B_q, \ldots)$ as in (10), p. 49, or (5) above, and $Y = [Y_p, Y_q, \ldots]$ where Y_p has p columns and n rows, Y_p has q columns and n rows, and so on. Then on expansion the commutantal equation breaks up into

$$BY_p = Y_pB_p,\ BY_q = Y_qB_q, \ldots$$

each of which can be solved separately. For brevity of statement let $p = 3$ which is typical, and let $Y_3 = [x, y, z]$ with n rows. Hence we must solve

$$B[x, y, z] = [x, y, z]\begin{bmatrix} d_1 & 1 & \cdot \\ d_2 & \cdot & 1 \\ d_3 & \cdot & \cdot \end{bmatrix}. \qquad \ldots \quad (8)$$

for x, y, z. Expansion gives

$$[Bx, By, Bz] = [d_1x + d_2y + d_3z, x, y],$$

whence $x = By$, $y = Bz$, and, by equating the first columns,

$$(B^3 - d_1B^2 - d_2B - d_3I)z = 0. \qquad \ldots \quad (9)$$

Let $\psi_p(B)$, $\psi_q(B)$, etc., denote the R.C.F.'s of B_p, B_q, etc., so that $\psi_p = \psi$ the R.C.F. of B, while each of ψ_p, ψ_q, ... contains its successor as a factor. Hence (8) is satisfied for an arbitrary z; and hence $Y_3 = [B^2z, Bz, z]$ is the general value of this commutant, where z, being arbitrary, has n parameters.

The general solution is given by combining the components Y_p, so that it is

$$\begin{aligned} Y &= [Y_p, Y_q, \ldots] \\ &= [B^{p-1}x, B^{p-2}x, \ldots, Bx, x, B^{q-1}y, B^{q-2}y, \ldots, By, y, \\ &\qquad\qquad B^{r-1}z, \ldots, z, \ldots] \end{aligned}$$

where $$\psi_p(B)x = 0, \ \psi_q(B)y = 0, \ \psi_r(B)z = 0, \ldots. \quad . \quad (10)$$

Subject to these last conditions the vectors $x, y, z, \ldots$ are arbitrary. Since ψ_q is a rational factor of ψ_p we may write $\psi_p = \phi_{p-q} \cdot \psi_q$, and likewise $\psi_q = \phi_{q-r}$, and so on, in terms of polynomials of orders $p - q$, $q - r$, etc., in B. The most general solutions for x, y, z, etc., are then given by $x = \xi$, $y = \phi_{p-q}(B)\eta$, $z = \phi_{p-r}(B)\zeta, \ldots$, where the ξ, η, ζ, etc., are entirely arbitrary column vectors; for such a substitution replaces the ψ's with different suffices $p, q, \ldots$ by ψ_p throughout in (10).

If the number of components $Y_p, Y_q, \ldots$ is k, there are apparently kn arbitrary constants among the vectors $\xi, \eta, \ldots$; but, owing to the various ranks of the premultipliers, $\phi_{p-q}(B)$, etc., these constants are fewer.

Now the rank of $\psi_p(B)$ is zero since this matrix vanishes identically, so that the n components of the vector x may be entirely arbitrary; and again, by example 2 below, the rank of $\psi_q(B)$ is $p - q$, so that y may have $n - p + q$ arbitrary constants (cf. p. 30); and the rank of $\psi_r(B)$ is $p+q-2r$, so that z may have $n-p-q+2r$ such constants. The next rank is $p + q + r - 3s$ and the vector may have $n - p - q - r + 3s$ components. And so on until all the equations (10) are utilized. Since $n = p + q + r + \ldots$ the total number of arbitrary constants among the vectors $x, y, \ldots$ is

$$\begin{aligned} n + (n - p + q) + (n - p - q + 2r) + (n - p - q - r + 3s) + \ldots \\ = p + 3q + 5r + 7s + \ldots \end{aligned}$$

where the coefficients ascend in arithmetical progression. This agrees with the total already found (cf. Ex. 2, p. 147), since $p = \sum_j p_{1j}$, $q = \sum_j p_{2j}$, and so on.

EXAMPLES

1. Show that the necessary and sufficient conditions for every non-zero vector in the field $\mathcal{F}$ to attain maximum grade with regard to an $n \times n$ matrix A are that $p = n$ and that the R.C.F. of A is irresoluble in $\mathcal{F}$ (cf. § 6, p. 53).

2. Prove that the ranks of the successive matrices $\psi_p(A)$, $\psi_q(A)$, . . . , when $\psi_p(\lambda)$, $\psi_q(\lambda)$, . . . are the successive invariant factors of A, are respectively 0, $p - q$, $p + q - 2r$, $p + q + r - 3s$, etc.

[A proof by the methods of § 7, p. 79, follows on using elementary divisors explicitly, e.g. if $C = \text{diag}\,(C_3(\alpha), C_2(\alpha), C_2(\beta), C_1(\beta))$ $\alpha \neq \beta$ (cf. p. 147), then the successive nullities of $(C - \alpha I)^m$ for $m = 0, 1, 2, \ldots$ are 0, 2, 4, 5, 5, . . . , and of $(C - \beta I)^{m'}$ are 0, 2, 3, 3, 3, ... Add the nullities to get that of $(C - \alpha I)^m (C - \beta I)^{m'}$: the rank is given by n less the nullity. A proof of this theorem by entirely rational methods would be desirable.]

3. If A has a single invariant factor, prove that its general commutant can be expressed as a polynomial in A of order $n - 1$.

[Here $p = n$, and $0 = q = r = \ldots$ so that, when $AX = XA$, the number of arbitrary constants in X is n. The polynomial $\Sigma c_i A^i$, $i = 0, 1, \ldots, p - 1$ satisfies the conditions.]

4. When $p = n$, solve the equation

$$[A^{n-1}x, A^{n-2}x, \ldots, Ax, x] = c_0 I + c_1 A + \ldots + c_{n-1} A^{n-1}$$

uniquely for the n components of x in terms of A and the arbitrary constants c_i.

[Reduce A to rational canonical form. If A is already in this form then $x = \{c_{n-1}, \ldots, c_1, c_0\}$.]

The Rational Solution of $AX = XB$.

The rational solution of $AX = XB$ proceeds in the same way. By iteration $A^2X = AXB = XB^2$, $A^rX = XB^r$, and $f(A)X = Xf(B)$ for any polynomial f. In particular, if ψ is the R.C.F. of B, then $\psi(B) = 0$, so that

$$\psi(A)X = 0.$$

For a non-zero X this cofactor must therefore be singular, and A must have one or more latent root in common with B. This means that $\phi(\lambda)$ and $\psi(\lambda)$, the R.C.F.'s of A and B respectively, have an H.C.F. $g(\lambda)$ of positive degree in λ; and this in turn enables ϕ and ψ to be factorized within the field $\mathcal{F}$, to which the elements of A and B belong. With the slightest modification the preceding methods apply. Thus, let $H^{-1}BH = \text{diag}\,(B_p, B_q, \ldots) = D$ be the usual rational canonical form of B, corresponding to the first, second, . . . invariant factors $\psi_p(\lambda)$, $\psi_q(\lambda)$, ... Then

$$AX = XHDH^{-1}, \text{ or } AY = YD \text{ where } Y = XH.$$

Put $Y = [Y_p, Y_q, \ldots]$; then $AY_p = Y_pB_p$, etc., whence, as before,

$$Y = [A^{p-1}x, A^{p-2}x, \ldots, Ax, x, A^{q-1}y, \ldots, y, \ldots],$$

where $x, y, \ldots$ are column vectors which must satisfy

$$\psi_p(A)x = 0, \quad \psi_q(A)y = 0, \quad \ldots,$$

and where the polynomials ψ refer to the R.C.F.'s of B. The number of arbitrary parameters is the sum of the nullities of $\psi_p(A), \psi_q(A), \ldots$. Finally $X = YH^{-1}$.

EXAMPLES

1. Show that the number of arbitrary parameters is equal to the sum of the degrees of the H.C.F.'s of pairs of the invariant factors, $\varphi_i(\lambda)$ and $\psi_j(\lambda)$ of B, taken in every way. (Frobenius, Cecioni.)

[A proof using elementary divisors is illustrated thus. Cf. Ex. 1, p. 148, where the invariant factors of A (or C) are $\varphi_4 = (\lambda - \alpha)^3(\lambda - \beta)$, $\varphi_2 = (\lambda - \alpha)^2$, and of B (or D) are $\psi_5 = (\lambda - \alpha)^2(\lambda - \beta)^2(\lambda - \gamma)$, $\psi_2 = (\lambda - \beta)^2$. Here $\psi_5(A)$, computed from the canonical form C, has nullity 5 and $\psi_2(A)$ has nullity 1; total 6. The commutant of $AX = XB$ is then

$$X = [A^4x, A^3x, A^2x, Ax, x, Ay, y]H,$$

where x has 5 and y has a single arbitrary constant. The degrees of the H.C.F.'s of (φ_4, ψ_5), (φ_4, ψ_2), (φ_2, ψ_5), (φ_2, ψ_2) are 3, 1, 2, 0 with a total 6. So also is the sum of the nullities of $\varphi_4(B)$, $\varphi_2(B)$.

2. Obtain a rational solution of this same equation $AX = XB$ in the form

$$X = K\{u, uB, uB^2, uB^3, v, vB\},$$

where the components $u, \ldots$ denote the rows of a six-rowed matrix, and where u has 4 and v has two arbitrary constants.

[Reduce A to rational canonical form C where $A = KCK^{-1}$, and consider the matrix $K^{-1}X$ row by row: u and v are its first and fifth rows.]

Triangular Form of Mutually Commuting Matrices.

When each pair of a set of $n \times n$ matrices $A, B, \ldots, D$ commutes it is possible to find a collineatory transformation H, H^{-1} which reduces them simultaneously to triangular form, by clearing the whole area that lies below their principal diagonals. To prove this we first note that by (9), p. 146, each of the matrices can be reduced by H to a diagonal of isolated submatrices, whose orders are fixed by the numbers of equal latent roots in any one of them, say A. (For the A of Ex. 1, p. 147, these orders are 5 and 3, corresponding to the two distinct roots α and β.) Each matrix of the set is now in the form diag $(D_1, D_2, \ldots, D_m)$, where corresponding submatrices must continue to commute and where all the latent roots of A_j (which belong to A) are equal, for each value of $j = 1, 2, \ldots, m$. Next, if the roots of another, B_j, are not all equal, a repetition of the process subdivides B_j and all corresponding submatrices D_j into isolated parts B_{jk} where the latent roots of B_{jk} are equal. Eventually after a finite number of such steps unequal latent roots are completely isolated for each of $A, B, \ldots, D$. Henceforward we assume that all the latent roots are equal for D_j, any member of the set. Also $A_jB_j = B_jA_j$ for any pair of the set.

Again, if the jth submatrices are of order 2×2 a further H transformation brings any one of them to the form $\begin{bmatrix} \lambda & \epsilon \\ \cdot & \lambda \end{bmatrix}$, where ϵ is either 0 or 1. If $\epsilon = 1$ in the case of any one of them, the rest must simultaneously be of the form $\begin{bmatrix} x & y \\ \cdot & x \end{bmatrix}$ in order to commute. And if, otherwise, $\epsilon = 0$ for each of them, then they are scalar. In either case the theorem is true when $n = 2$.

It now follows in general for any n by induction on assuming its truth for lower values. This is done by throwing A_j into classical canonical form according to the Weyr characteristic (p. 80); that is, the rows and columns of $\operatorname{diag}(C_{p_1}(a), C_{p_2}(a), \ldots)$ being numbered $1, 2, \ldots, p_1, \ldots, p_1 + p_2, \ldots$, they are now to be rearranged in the order

$$1,\ p_1 + 1,\ p_1 + p_2 + 1, \ldots,\ 2,\ p_1 + 2, \ldots,\ 3, \ldots$$

The effect of this is to express A_j in the form $aI + U^q$, where q is the number of its elementary divisors, and the broken sequence of units upon the over-diagonal is replaced by a continuous sequence upon the qth over-diagonal, as in

$$\operatorname{diag}(C_3(a), C_2(a)) = \left[\begin{array}{ccc:cc} a & 1 & \cdot & \cdot & \cdot \\ & a & 1 & \cdot & \cdot \\ & & a & \cdot & \cdot \\ \hdashline & & & a & 1 \\ & & & & a \end{array}\right] = K \left[\begin{array}{cc:cc:c} a & \cdot & 1 & \cdot & \cdot \\ & a & \cdot & 1 & \cdot \\ \hdashline & & a & \cdot & 1 \\ & & & a & \cdot \\ \hdashline & & & & a \end{array}\right] K^{-1},$$

where K is the requisite permutation matrix which deranges 1 2 3 4 5 to 1 4 2 5 3. The significance of the dotted partitions is explained on representing the Segre characteristic $\{3, 2\}$ by the graph $\begin{array}{ccc} * & * & * \\ * & * & \end{array}$ which is read in the order $\begin{array}{ccc} 1 & 2 & 3 \\ 4 & 5 & \end{array}$ and by transposing it to $\begin{array}{cc} 1 & 4 \\ 2 & 5 \\ 3 & \end{array}$ which exhibits the Weyr characteristic $[2, 2, 1]$.

When this is done, then any B_j that commutes with A_j is brought to the same Weyr partitioned form, say, $T(B_{q_iq_j})$, where $[q_1, q_2, \ldots]$ is the partition conjugate to $\{p_1, p_2, \ldots\}$; and, in this, all below the $B_{q_jq_j}$ upon the principal diagonal is now zero (the z_0, z_1 of Ex. 1, p. 147, now reappear above the diagonal $x_0t_0x_0t_0x_0$ of deranged x_0, t_0). If no pair of p_j are equal the reduction is complete, as in the cited example, but if through equalities elements are left within $B_{q_jq_j}$, but below the diagonal of α's, then a further H transformation removes them since the theorem is assumed true when $q_j < n$. The only remaining case is when $q_1 = n$, which means that the Segre characteristic is $\{1\ 1\ 1 \ldots\}$ and the matrix is scalar. If any one of the original set is not scalar, take it as A with $q_j < n$; otherwise all are scalar. In either case the theorem is true.

EXAMPLE

Prove that, when A and B commute, each latent root of AB is of the form $\lambda\mu$, where λ and μ are latent roots of A and B respectively. Generalize this result. (Frobenius.)

[Use triangular canonical forms: $f(\lambda, \mu, \ldots)$ is a latent root of $f(A, B, \ldots)$.

Historical Note.

G. Frobenius treated commutantal matrices in the *Berliner Sitzungsberichte* (1896), 601–614, as also I. Schur, *op. cit.* (1902), 120–125.

Recent work upon commutantal equations $AX = XA$, $AX = XB$ treated by rational methods will be found in papers by D. E. Rutherford, *Akad. Wetensch. Amsterdam Proc.*, **35** (1932), 54–59, and *Proc. Royal Soc. Edinburgh*, **62** (1949), 454–459. Rutherford stresses the importance of the Weyr characteristic; and upon his treatment much of the above exposition is founded. See also an explicit rational solution of $AX = XB$ by H. O. Foulkes, *Proc. London Math. Soc.*, **2**, 50 (1949), 196–209, and a general treatment by M. P. Drazin, J. W. Dungey and K. W. Gruenberg, *Journal London Math. Soc.*, **26** (1951), 221–228, and the latter in *Proc. London Math. Soc.* (3), **1** (1951), 222–231. Many references are to be found in the work by C. C. MacDuffee, *The Theory of Matrices* (New York, 1946), while his *Vectors and Matrices*, Carus Mathematical Monograph **7** (1943), contains an exposition of the factorized rational canonical form of a matrix.

MISCELLANEOUS EXAMPLES

1. Find the elementary divisors of the second compound of the matrix $\lambda I + U$, for orders 4 and 5 respectively.

2. Find the elementary divisors of the third compound of the classical canonical matrix $[C_4(\alpha), C_1(\beta)]$.

3. A matrix of elements which are functions of a variable x will be said to be differentiated when each element is differentiated. Writing

$$A_{(x)} \equiv \left[\frac{d}{dx} a_{ij}\right],$$

prove the rule for differentiating a product of matrices, namely

$$(AB)_{(x)} = A_{(x)}B + AB_{(x)}.$$

Deduce the result

$$(A^{-1})_{(x)} = -A^{-1}A_{(x)}A^{-1}, \quad |A| \neq 0.$$

4. If $A \equiv [a_{ij}], B, \ldots, K$ are a finite number of matrices of order n, and if $\hat{A}, \hat{B}, \ldots, \hat{K}$ are a set of the same order, permutable with each other and such that

$$A\hat{A} + B\hat{B} + \ldots + K\hat{K} = 0,$$

prove that the matrices $\hat{A}, \hat{B}, \ldots, \hat{K}$ jointly satisfy a characteristic equation of the nth order, namely,

$$|(a_{ij}\hat{A} + b_{ij}\hat{B} + \ldots + k_{ij}\hat{K})| = 0. \qquad \text{(H. B. Phillips.)}$$

[The Cayley-Hamilton equation corresponds to the case,

$$AI - IA = 0.]$$

5. Find the latent roots of a given matrix B which commutes with a given A.

[Reduce A to classical form HAH^{-1}. The elements on the leading diagonal of the corresponding HBH^{-1} are the latent roots of B.]

6. If $A, B, \ldots, K$ are permutable matrices of order n, with latent roots $\alpha_i, \beta_i, \ldots, \varkappa_i$, $i = 1, 2, \ldots, n$, prove that these roots may be taken in such an order for each matrix that the latent roots of a matrix polynomial $f(A, B, \ldots, K)$ are $f(\alpha_i, \beta_i, \ldots, \varkappa_i)$. (Frobenius, Phillips.)

[Use Ex. 5.]

7. The set of roots $\alpha_i, \beta_i, \ldots, \varkappa_i$ in the preceding example will be called a set of *corresponding* roots. Prove that if a matrix X is permutable with a matrix A, an analytic function $f(X)$ can be expanded in a convergent Taylor series,

$$f(X) = f(A) + (X - A)f'(A) + \ldots + \frac{(X - A)^r}{r!} f^{(r)}(A) + \ldots,$$

provided that each latent root of X lies within a circle with centre at the corresponding root of A, within which the scalar function $f(z)$ is convergent. (Phillips.)

8. To check a product of numerical matrices.

(i) Two matrices. $AB = [a_{ij}][b_{ij}] = [c_{ij}] = C.$

Add the elements in each *column* of A, obtaining a row vector $[a_1, a_2, \ldots, a_n] = a'$. Add the elements in the *rows* of B, obtaining a column vector b. Then the scalar $a'b$ is the sum of *all* the elements of the product C.

(iii) Three matrices. $APB = C.$

The scalar $a'Pb$ is the sum of the elements in C. (W. E. Roth.)

[The matrices may of course be rectangular.]

9. Prove that the matrix

$$Q = \tfrac{1}{2}\begin{bmatrix} -1 & 1 & 1 & 1 \\ 1 & -1 & 1 & 1 \\ 1 & 1 & -1 & 1 \\ 1 & 1 & 1 & -1 \end{bmatrix}$$

is orthogonal.

10. The compounds of a unitary matrix are also unitary. (Rados.) [Cf. *Invariants*, p. 165.]

11. If A is an orthogonal matrix, the pencil $\lambda A - \mu A'$ is equivalent to the pencil $\lambda A^2 - \mu I$. Hence the invariant factors of the pencil are those of A^2, made homogeneous.

12. If A is a square matrix there exist unitary matrices P and Q such that

$$PAQ = [\alpha_i \delta_{ij}]. \qquad \text{(C. Jordan, Autonne.)}$$

[Multiply the above by its transposed conjugate. We deduce that the α_i are square roots of the latent roots of the Hermitian matrix $\bar{A}'A$.]

13. If $X = AB$ and $Y = BA$, prove that $Xf(X) = Af(Y)B$.

14. Prove that the matrices $X = AB$ and $Y = BA$ always have the same latent roots. If $\psi(\lambda)$, $\chi(\lambda)$ are the reduced characteristic polynomials corresponding to X and Y, prove that either $\psi(\lambda) = \chi(\lambda)$, or $\lambda\psi(\lambda) = \chi(\lambda)$, or $\psi(\lambda) = \lambda\chi(\lambda)$.

[Show by Ex. 13 that $X\chi(X) = 0 = Y\psi(Y)$. Then write $\psi(X) = X^r\psi_1(X)$, where $\psi_1(0) \neq 0$; similarly for $\chi(Y) = Y^s\chi_1(Y)$. Prove $r + 1 \geqslant s$, $s + 1 \geqslant r$.]

15. If $X = [x, y, \ldots, z]$ is a matrix of ω column-vectors $x, y, \ldots, z$ each consisting of n elements, show that $M = X\tilde{X}'$, $N = \tilde{X}'X$ are square matrices of orders n and ω respectively, which possess the following properties.

(i) They are non-negative definite (except in the complex symmetrical case).

(ii) Their R.C.F.'s are the same or else differ by a single factor M or N.

(iii) If either is orthogonal (or unitary), so is the other.

[(i) $\tilde{\xi}'M\xi = \sum_x^z \tilde{\xi}'x\tilde{x}'\xi = \Sigma(\tilde{\xi}'x)(\xi'\tilde{x}) \geqslant 0$. Cf. p. 97. For N, treat X as a set of n row-vectors.]

16. Show that every symmetric matrix $A + A'$ can be resolved (i) into rational factors $H'DH$, where D is diagonal; and (ii) into a pair of factors $K'K$, which are not necessarily rational in the elements of A.

[(i) p. 85: (ii) $K = \sqrt{D}.H.$]

17. If $H = \begin{bmatrix} -q & 1 \\ 1 & . \end{bmatrix}$, $U' = \begin{bmatrix} . & . \\ 1 & . \end{bmatrix}$, $R = \begin{bmatrix} . & 1 \\ p & q \end{bmatrix}$, evaluate the product of six-row matrices

$$\Gamma\Delta \equiv \begin{bmatrix} & & H \\ & H & \\ H & & \end{bmatrix} \begin{bmatrix} R & U' & \\ & R & U' \\ & & R \end{bmatrix} = \begin{bmatrix} . & . & HR \\ . & HR & HU' \\ HR & HU' & . \end{bmatrix},$$

showing that the result is symmetrical.

18. Determine a general real canonical form for a pencil of real quadrics.

[Using the real classical canonical form of p. 72 and the above example, proceed as on p. 132. The new feature is the canonical submatrix of type $\Gamma(\Delta - \lambda I)$, corresponding to an elementary divisor of type $(\lambda^2 - q\lambda - p)^e$. In Ex. 17, $e = 3$.]

19. Find every quadric $x'Rx$ which reciprocates a given non-singular quadric $x'Ax$ into a given non-singular tangential quadric $u\Gamma u'$, $(R = R', A = A', \Gamma = \Gamma')$.

[First show that $R\Gamma R = A$; whence $R\Gamma = \sqrt{A\Gamma} = f(A\Gamma)$ where f is a polynomial. Verify that $f(A\Gamma)\,\Gamma^{-1}$ is *symmetric*.]

20. Discuss the problem of determining a matrix X which reciprocates a given A into a given B, when one or more of these three matrices are symmetric, skew, Hermitian, or general.

[Use the methods of Ex. 16 and 19.]

21. By considering the Segre characteristic show that there are seven distinct types of pencils of conics (including one singular pencil).

Identify the characteristics in the case of a pair of base conics with single contact, double contact, three-point contact, four-point contact.

$$\left[[2, 1], \begin{bmatrix} 1, & 1 \\ 1 & \end{bmatrix}, [3], \begin{bmatrix} 2 \\ 1 \end{bmatrix}. \right]$$

22. Discuss the corresponding problem of a pencil of quadric surfaces.

[Fourteen non-singular, and one singular case.]

23. If α_x denote the continued fraction

$$\frac{1}{a_{11} - x -}\,\frac{a_{12}}{|a_{22} - x -}\,\frac{a_{23}}{|a_{33} - x -}\cdots\frac{a_{n-1,\,n}}{|a_{nn} - x},$$

then $\left(\dfrac{d}{dx}\right)^m \alpha_x$, where m is a positive integer, is the leading element in the matrix $m!(A - xI)^{-m-1}$, where A denotes the continuant matrix with the a_{ij}'s indicated and $a_{i+1,\,i} = 1$, corresponding to α_x. (Whittaker.)

[*Note:*—The integral of the continued fraction of the preceding example is the leading element in the matrix $-\log(A - xI)$.]

24. From n linear functions ξ_i defined by

$$\xi_i = a_{i1}x_1 + a_{i2}x_2 + \ldots + a_{in}x_n, \quad \text{or} \quad \xi = Ax,$$

all the powers and products of degree m are formed, and are taken as the components of a vector $\xi^{[m]}$. If this vector is related to the corresponding vector $x^{[m]}$ by the equation

$$\xi^{[m]} = A^{[m]}x^{[m]},$$

prove that for such matrices

$$[AB]^{[m]} = A^{[m]}B^{[m]},$$
$$[A^{-1}]^{[m]} = [A^{[m]}]^{-1}, \quad |A| \neq 0.$$

Deduce that the latent roots of $A^{[m]}$ are the m-ary powers and products of the latent roots of A. (Franklin.)

25. Let $A^{[m]}$ be modified by having each column divided by the square root of the multinomial coefficient in its first element, and each corresponding row multiplied by the square root. Prove that if A is Hermitian or unitary, the modified $A^{[m]}$ is Hermitian or unitary.

26. Find the elementary divisors of $[C_3(\alpha)]^{[2]}$, $[C_4(\alpha)]^{[2]}$.

27. Solve the following equation in x, after determining the conditions under it is soluble:

$$\begin{bmatrix} \lambda & x & & \\ & \lambda & x & \\ & & \lambda & x \\ & & & \lambda \end{bmatrix} H = H \begin{bmatrix} \lambda & a_1 & a_2 & a_3 \\ & \lambda & b_1 & b_2 \\ & & \lambda & c_1 \\ & & & \lambda \end{bmatrix}, \quad |H| \neq 0.$$

28. The value of the multiple integral

$$\int (x'x)\,dx, \quad \text{over the range } x'Ax \leqslant 1,$$

where the form $x'Ax$ is positive definite, is the sum of the diagonal elements in A^{-1} multiplied by $\frac{1}{2}\pi^{\frac{1}{2}n}A^{-\frac{1}{2}}/\Gamma(\frac{1}{2}n+2)$.

29. The maximum and minimum values of $q \equiv x'Ax$, subject to the conditions

$$x'Bx = p, \quad Cx = 0, \quad C \text{ being of order } (n-2) \times n,$$

are given by the equation, in a quadripartite determinant,

$$\begin{vmatrix} Bq - pA, & C \\ C & . \end{vmatrix} = 0.$$

30. The value of the multiple integral

$$(2\pi)^{-\frac{1}{2}n} \int_{-\infty}^{\infty} (\tfrac{1}{2}x'P_1x)^{m_1} (\tfrac{1}{2}x'P_2x)^{m_2} \ldots (\tfrac{1}{2}x'P_kx)^{m_k} \exp(-\tfrac{1}{2}x'Ax)\,dx$$

is the coefficient of

$$\lambda_1^{m_1}\lambda_2^{m_2}\ldots\lambda_k^{m_k}/(m_1!\; m_2!\ldots m_k!)$$

in the expansion of the determinant

$$|A - \lambda_1P_1 - \lambda_2P_2 - \ldots - \lambda_kP_k|^{-\frac{1}{2}}.$$

31. If $A = [a_{ij}]$ is a matrix of complex elements, and $\breve{a}_{ij}$ denotes the modulus of a_{ij}, prove the inequality, for all elements,

$$\overbrace{(AB)} \leqslant \breve{A}.\breve{B}.$$
$$\underbrace{(A+B)} \leqslant \breve{A} + \breve{B}.$$

32. If s_i is the sum of the absolute values of the elements in the ith row of a complex matrix A, and t_j is the corresponding sum for the jth column, and s and t are the greatest of these, then the latent roots λ of A are such that

$$|\lambda| \leqslant \tfrac{1}{2}(s+t). \qquad \text{(Browne.)}$$

33. Deduce, from the above, the inequality for latent roots of an Hermitian matrix, and for α and β in the latent roots $\alpha + i\beta$ of a general matrix. (Browne.)

34. The function of frequency (or probability) for two correlated variables x and y is

$$\varphi(x, y) = \sigma_1^{-1}\sigma_2^{-1}(2\pi)^{-1}(1-\rho^2)^{-\frac{1}{2}}\exp-\left(\frac{1}{2(1-\rho^2)}\left(\frac{x^2}{\sigma_1^2} - 2\rho\frac{xy}{\sigma_1\sigma_2} + \frac{y^2}{\sigma_2^2}\right)\right).$$

If N samples of paired measures $\{x_i, y_i\}$ are taken, and the three quadratic moment coefficients are calculated about the mean $\{h, k\}$, where $N\{h, k\} = \sum_N \{x_i, y_i\}$, by the formulæ

$$m_{11} = \frac{1}{N}\Sigma(x_i - h)^2 - h^2, \quad m_{12} = \frac{1}{N}\Sigma(x_i - h)(y_i - k) - hk,$$

$$m_{22} = \frac{1}{N}\Sigma(y_i - k)^2 - k^2,$$

prove that the mean value of the vector $\{m_{11}, m_{12}, m_{22}\}$, in all possible samples of N, is given by $\frac{N-1}{N}\{\sigma_1^2, \rho\sigma_1\sigma_2, \sigma_2^2\}$.

[The mean value will involve a $2N$-ple integral.]

35. If $E = [e_{ij}]$ is a matrix of order n, where $e_{ij} = 1$ for all i and j, prove that the matrix $I - \frac{1}{n}E$ is of rank $n - 1$.

[The above may be proved directly. It can also be deduced from the fact that the sum of the squared deviations of n numbers from their arithmetic mean is reducible to a sum of $n - 1$ independent squares; for the fact that the sum of the deviations themselves must be zero imposes one relation of linear dependence on the variables.]

36. Prove that the sum of the $\frac{1}{2}n(n-1)$ squared differences of n numbers taken in pairs is a quadratic form of rank $n - 1$.

[It will be found to be the same quadratic form, apart from a constant factor, as the sum of squared deviations about the mean in the previous example. Cf. A. L. Bowley, *Elements of Statistics*, pp. 114–5.]

37. Transform the multiple integral

$$\int_{\{-\infty\}}^{\{\infty\}} \varphi(x'Ax)dx \quad \text{into the form} \quad c\int_0^\infty \psi(\chi)d\chi$$

by a spherical-polar transformation, where $\chi = x'Ax$ is a positive definite quadratic form in n variables.

[First reduce to a sum of squares by a congruent transformation.]

38. If

$$R(\lambda) \equiv \sum_{k=1}^{n} \frac{R_k}{\lambda - \alpha_k}$$

is the resolvent of a square matrix $A = [a_{ij}]$, all whose latent roots are distinct, prove that the numerators R_k of the partial fractions can be written in the form

$$R_k = \left[\frac{\partial\alpha_k}{\partial a_{ji}}\right] \equiv \Omega\alpha_k$$

where Ω is the matrix differential operator whose ijth element is $\partial/\partial a_{ji}$.

39. A real orthogonal matrix P exists which reduces any given real matrix A to triangular form $[\Gamma_{ij}]$, where each submatrix Γ is real and of orders 2×2, and where

$$\Gamma_{ii} = \begin{bmatrix} \alpha & \beta \\ -\beta & \gamma \end{bmatrix}, \quad \Gamma_{ij} = 0 \text{ when } i > j.$$

(F. D. Murnaghan and A. Wintner.)

[By the methods of Ex. 14, p. 109, form an identity

$$A\,[a, b] = [a, b] \begin{bmatrix} \mu & \nu \\ -\nu & \mu \end{bmatrix},$$

where $a + ib$ is a latent vector corresponding to a latent root $\mu + i\nu$. Using $a + ib \neq 0$, show that a and b are both non-zero and linearly independent. Refer the matrices and vectors to an orthogonal frame B, consisting of the internal and external bisectors of the angle made by the vectors a and b, together with $n - 2$ further orthogonal axes. Proceed by the methods of the cited example.]

40. If A is an arbitrary non-singular matrix, then the matrix $Q = \sqrt{(\tilde{A}'A)}\,.\,A^{-1}$ exists, and is unitary (or orthogonal): so also is $A^{-1}\sqrt{(A\tilde{A}')}$.

[If $\tilde{A}'A = B = P^2$, then P exists as a polynomial in B (by (30), p. 78): also $\tilde{P}' = P$. Further, if $Q = PA^{-1}$, then $\tilde{Q}'Q = I$.]

41. Any non-singular matrix A may be resolved into a pair of factors $R_1\Theta_1$, and also into $\Theta_2 R_2$, where the R factor is Hermitian (or symmetric), and the Θ factor is unitary (or orthogonal). (F. D. Murnaghan and A. Wintner.)

[In Ex. 40 write $A = Q^{-1}P = \Theta_2 R_2$. Similarly for $R_1\Theta_1$.]

42. The resolution of a non-singular matrix A into $R_1\Theta_1$ and also into $\Theta_2 R_2$ is *unique* when R_1 and R_2 are each restricted to be positive definite.

Justify the nomenclature: *left- and right-handed polar co-ordinates* (R_1, Θ_1), (Θ_2, R_2) of a non-singular matrix A. Show that these alternative modes coalesce if and only if A commutes with $\tilde{A}'$.

(F. D. Murnaghan and A. Wintner.)

[By considering the latent roots of $\sqrt{B}$, taking exactly ν latent roots of B to be distinct, show that $\sqrt{B}$ has 2^ν possible alternative forms, only one of which is positive definite.

The scalar formula $z = r(\cos\theta + i\sin\theta)$ is a particular case of the matrix formula when A is a matrix of order unity.

Only if A commutes with $\tilde{A}'$ will $R_1 = R_2$ and $\Theta_1 = \Theta_2$. In this case A is sometimes termed a normal matrix.]

43. Let $H = x\bar{y}' + y\bar{x}'$, where x and y are column vectors of n complex elements. Prove that

$$H^3 - (\bar{\gamma} + \gamma)\,H^2 - (\alpha\beta - \bar{\gamma}\gamma)\,H = 0,$$

where $\bar{x}'x = \alpha,\ \bar{y}'y = \beta,\ \bar{x}'y = \gamma.$

Obtain the corresponding equations satisfied by

$$Q = xy' + yx',\ S = xy' - yx'.$$

(L. S. Goddard)

INDEX

A series of hardcover reprints of major works in mathematics, science and engineering.
All editions are 5⅝ × 8½ unless otherwise noted.

Mathematics

Theory of Approximation, N. I Achieser. Unabridged republication of the 1956 edition. 320pp. 49543-4

The Origins of the Infinitesimal Calculus, Margaret E. Baron. Unabridged republication of the 1969 edition. 320pp. 49544-2

A Treatise on the Calculus of Finite Differences, George Boole. Unabridged republication of the 2nd and last revised edition. 352pp. 49523-X

Space and Time, Emile Borel. Unabridged republication of the 1926 edition. 15 figures. 256pp. 49545-0

An Elementary Treatise on Fourier's Series, William Elwood Byerly. Unabridged republication of the 1893 edition. 304pp. 49546-9

Substance and Function & Einstein's Theory of Relativity, Ernst Cassirer. Unabridged republication of the 1923 double volume. 480pp. 49547-7

A History of Geometrical Methods, Julian Lowell Coolidge. Unabridged republication of the 1940 first edition. 13 figures. 480pp. 49524-8

Linear Groups with an Exposition of Galois Field Theory, Leonard Eugene Dickson. Unabridged republication of the 1901 edition. 336pp. 49548-5

Continuous Groups of Transformations, Luther Pfahler Eisenhart. Unabridged republication of the 1933 first edition. 320pp. 49525-6

Transcendental and Algebraic Numbers, A. O. Gelfond. Unabridged republication of the 1960 edition. 208pp. 49526-4

Lectures on Cauchy's Problem in Linear Partial Differential Equations, Jacques Hadamard. Unabridged reprint of the 1923 edition. 320pp. 49549-3

The Theory of Branching Processes, Theodore E. Harris. Unabridged, corrected republication of the 1963 edition. xiv+230pp. 49508-6

The Continuum, Edward V. Huntington. Unabridged republication of the 1917 edition. 4 figures. 96pp. 49550-7

Lectures on Ordinary Differential Equations, Witold Hurewicz. Unabridged republication of the 1958 edition. xvii+122pp. 49510-8

Mathematical Methods and Theory in Games, Programming, and Economics: Two Volumes Bound as One, Samuel Karlin. Unabridged republication of the 1959 edition. 848pp. 49527-2

Famous Problems of Elementary Geometry, Felix Klein. Unabridged reprint of the 1930 second edition, revised and enlarged. 112pp. 49551-5

Lectures on the Icosahedron, Felix Klein. Unabridged republication of the 2nd revised edition, 1913. 304pp. 49528-0

On Riemann's Theory of Algebraic Functions, Felix Klein. Unabridged republication of the 1893 edition. 43 figures. 96pp. 49552-3

A Treatise on the Theory of Determinants, Thomas Muir. Unabridged republication of the revised 1933 edition. 784pp. 49553-1

A Survey of Minimal Surfaces, Robert Osserman. Corrected and enlarged republication of the work first published in 1969. 224pp. 49514-0

The Variational Theory of Geodesics, M. M. Postnikov. Unabridged republication of the 1967 edition. 208pp. 49529-9